FOUNDATION ENGINEERING:
DESIGN AND CONSTRUCTION IN TROPICAL SOILS

BALKEMA – Proceedings and Monographs
in Engineering, Water and Earth Sciences

Foundation Engineering: Design and Construction in Tropical Soils

Edited by

Bujang B.K. Huat
University Putra Malaysia, Malaysia

Faisal Haji Ali
University of Malaya, Malaysia

Husaini Omar
University Putra Malaysia, Malaysia

Harwant Singh
University Malaysia Sarawak, Malaysia

Taylor & Francis
Taylor & Francis Group

LONDON/LEIDEN/NEW YORK/PHILADELPHIA/SINGAPORE

Published by: A.A. Balkema Publishers, a member of Taylor & Francis Group plc
P.O. Box 447, 2300 AK Leiden, The Netherlands
e-mail: Pub.NL@tandf.co.uk
www.balkema.nl, www.tandf.co.uk, www.crcpress.com

ISBN 0-415-39898-3

TABLE OF CONTENTS

Preface

Almost all structures constructed on earth will transmit their load to the earth through their foundation. In other words, foundation is the lowest part of a structure, which is in contact with the soil and transmits the load of the structure to the ground. Therefore the design of the foundation is very important to ensure the stability of the structure it supports. Foundations are generally divided into two categories – shallow foundation and deep foundation.

A shallow foundation is foundation built near or at ground (soil) or rock surface. This foundation is placed on firm soil near the ground and beneath the lowest part of the superstructure. Examples of these foundations are pad footing and spread footing. Deep foundation on the other hand is foundation that transmits structural loads to deeper soils or rock layers that are far from the surface. This foundation is constructed on a soil that is not firm, and transmits the load of the structure considerably below the ground of the lowest part of the superstructure. Deep foundations can be divided into two categories, that is pile foundation and drilled pier foundation or caisson. Piles are normally columns made of concrete, wood, plastic or steel that are driven into the ground. Drilled pier or caisson on the other hand is a special pile made of cast *in situ* concrete inside a bored or drilled hole.

Design methods for a pile foundation can generally be regarded as a mixture of rational and empirical techniques, and tend to vary between geographical regions, partly from the instructor's influence and partly because there are few "design absolutes" since we do not fully understand the behavior of a foundation. Rational techniques are those developed from the principles of physics and science, and are useful ways to describe mechanisms we understand and are able to quantify. Conversely empirical techniques are based on experimental data and local experiences on physical mechanisms where we have a limited understanding only.

In a developing tropical country such as Malaysia, documented works on local foundation engineering practices to date, can best be described as fragmented. For a country that has enjoyed rapid infrastructure development over the last two decades, it is felt necessary to compile these experiences in a more orderly manner for the benefits of the newer generation and practicing engineers as well as students. The outcome of such efforts is this book. We envisage this book to be a valuable reference not only for tropical Malaysia but also for other countries with similar geology and climate. One of the keys to successful foundation engineering is to understand the mix between rationalism and empiricism, the strengths and limitations of each, and how to apply them to practical design problems.

Bujang B.K. Huat
Faisal Haji Ali
Husaini Omar
Harwant Singh

Acknowledgements

We are very grateful to a good many people who have contributed to the realization of this book. Many meetings had to be called and attended, countless hours spend on writing and rewriting the chapters, and numerous letters and email messages sent and received. The organization of all these is attributed to several individuals to whom we wish to place on record our appreciation.

We would like to extend our thanks to all the authors and co-authors, as well as other contributors, and to our language editor Ms. Sumangala Pillai of University Putra Malaysia Press.

Last but not least we would like to express our sincere gratitude to Ms. Ernaleeza Mahsum and Ms. Norzuwana Wahab who are largely responsible for the technical formatting of the book.

Editors

CHAPTER 1

Foundation Engineering in Tropical Soils

Bujang B.K. Huat
Department of Civil Engineering, University Putra Malaysia, Malaysia

Faisal Hj. Ali
Department of Civil Engineering, University of Malaya, Malaysia

1.1 BRIEF HISTORY OF FOUNDATION ENGINEERING

Almost all structures constructed on earth will transmit their loads to the earth through their foundations. Essentially, the foundation is the lowest part of a structure that is in contact with the soil and transmits the load of the structure to the ground. Therefore the design of the foundation is very important to ensure the stability of the structure it supports. Although the importance of building a good foundation for any construction has been recognized for thousands of years, the discipline of foundation engineering, as we know it today, was not developed until about the early nineteenth century. The early foundation design was mainly based on general knowledge and past experience. For example, stone masonries were built in the New York City supported by foundation made of compacted gravel and of dimension 1.5 times the width of the wall.

Eiffel tower in Paris (Figure 1.1) is a good example of a new structure based on the principle of "modern" foundation engineering. Alexander Gustave Eiffel built the tower in 1889. Eiffel realized that a good foundation is vital for the tower in order not to suffer the same fate as the Leaning

(a) (b)

Figure 1.1. (a) Eiffel tower in Paris (*http://www.intermac.co.uk/homepage/photographs/placed/paris/effieltower.jpg*). (b) Leaning tower of Pisa (*http://hwlaartgallery.com/photo%20Album.htm*).

tower of Pisa in Italy. Before constructing the tower, Eiffel had invented a "new" method of soil exploration. The method included driving into soil 200 mm diameter pipe filled with compressed air. The air prevented ground water from entering the tube thus enabling high quality soil samples to be obtained. Based on this study of in situ soil condition, Eiffel had successfully identified and thereby avoided areas of soft ground that could cause problems to the tower. He was therefore able to place all four feet of the tower on suitable foundation on firm ground.

Today, our knowledge of design and construction of foundations is far more advanced than it was hundreds of years ago. We can now design foundations that are reliable, cost saving and of high capacity for various types of modern structures.

Modern foundation engineering can be said to comprise various disciplines. This would include structural engineering, geotechnical engineering and construction engineering.

1.2 FOUNDATION ENGINEERING AND TROPICAL SOILS

Design methods for a modern foundation can generally be regarded as a mixture of rational and empirical techniques, and tend to vary between geographical regions, partly from instructor's influence and partly because there are few "design absolutes". This is because we still do not fully understand the behavior of a foundation. The rational technique was developed from the principles of physics and science, and is useful to describe mechanisms we understand and are able to quantify. Conversely, the empirical technique is based on experimental data and local experience of physical mechanisms, of which we only have a limited understanding.

Most of the theories used in our foundation design tend to be those that have been developed based on the experience of construction on temperate zone soils, which mostly consist of transported (sedimentary) type of soils. But in tropical countries where soils can be literally classified as tropical soils, the mode of formation and hence the physical properties of the soils somewhat differ. This would inevitably affect the way we design and construct our foundation. This is because a key to successful foundation engineering is to understand the mix between rationalism and empiricism, the strengths and limitations of each, and how to apply them to practical design problems.

Areas with tropical climates (the *A category* climate in the Köppen Classification System) are extensive, occupying almost all of the continents between latitudes 20°N to 20°S of the equator as illustrated in Figure 1.2.

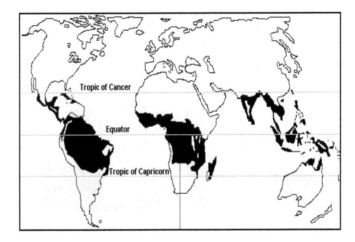

Figure 1.2. Areas with a tropical climate. (*Source*: World Wide Web, http://www.nctwr.org.au/image/tropic-map.jpg).

The key criterion of an *A category* climate is for the coolest month to have a temperature of more than 18°C making it the only true winterless climate category of the world. The consistent day lengths and almost perpendicular sun angle throughout the year generates temperatures above 18°C. Another characteristic is the prevalence of moisture. Warm, moist and unstable air masses frequent the oceans at these latitudes. As a consequence, this climate zone has abundant sources of moisture giving rise to high humidity.

Due to this climatic condition, weathering of parent rocks (igneous, sedimentary or metamorphic), mainly chemical weathering, is the main agent for soil formation in the tropics. The soils formed by weathering are largely left in place, thereby literally called residual soil, and whose character depends on the parent rock it develops from. For example, residual soil on weathered granite will initially be sandy, as sand-sized particles of quartz and partially weathered feldspar are released from the granite. The partially weathered feldspar grains will gradually over time further completely weather into fined-grained clay minerals. As the resistant quartz does not weather, the resulting soil will have both sand-sized quartz and clay. This will further change over time as this residual soil that develops from granite may become more clayey. However, the influence of the parent rock decreases over the passage of time. After a sufficient time period, the differences in the residual soils from different types of rocks i.e. igneous, sedimentary and metamorphic may be obliterated. The present or absence of coarse grains quartz in the parent rock becomes the only vestige that survives and has a long-term significance.

In addition to the influence of parent rocks, residual soils in the tropics, have a vertical soil section, called the soil profile, which consists of a distinct layering termed the soil horizons formed more or less parallel to the ground surface. These genetically related horizons are a reflection of the weathering process. The soil profile also has a weathering aspect that gives rise to a vertical weathering profile that is a critical aspect from the engineering perspective. The weathering profile reflects the state of weathering along the soil profile or vertical soil section from the bedrock (unaltered parent rock) to the ground surface. It consists of material that shows progressive stages of transformation or "grading" from fresh rock to completely weathered material towards the ground surface. The weathering profile portrays considerable variation from place to place due to the local variation in rock type and structure, topography and rates of erosion because of regional climatic variation, particularly rainfall. The entire weathering profile, generally, indicates a gradual change from fresh rock to a completely weathered soil as illustrated in Figure 1.3.

Larger expanses of residual soils with greater depths, due to active weathering leading to residual soil formation, are normally found in tropical humid regions, such as Northern Brazil, Ghana, Malaysia, Nigeria, Southern India, Sri Lanka, Singapore and the Philippines. The depth of these soils may extend to several hundred meters.

In addition to the residual soils, transported soils are also found in the tropics, though in terms of coverage, but not necessarily in the order of significance, they are less in extent than the residual soils. Transported soils by definition are soils that are formed from materials formed elsewhere and which have moved to the present site where they constitute the unconsolidated superficial layer. The physical processes through the operation of their agents of transportation i.e. mainly gravity and water have dislodged, eroded and transported soil particles to their present location.

Figure 1.4 shows an example of a simplified soil map of Malaysia, one of the countries in the tropics. As shown, about 70% of the country's land area is covered by residual type soils, while the remaining 30% is marked as alluvial, and falls under the category of transported soils. By virtue of their location (they are mostly found along the coast), these soils are called the coastal alluvium. Alluvial soils, which are also known as fluvial soils or alluvium, are soils that are transported by rivers. Marine soils are also deposited in water, but by salt water. These soils are normally found in engineering because many engineering structures are actually built on them. They often fall into the category of soft soils because of their low strength and high compressibility.

Another type of soils found in the tropics is the organic soils. By definition, organic soils are soils with more than 20% organic matter while peat is organic soil with more than 75% organic matter. Other components of soils may include sand, silt and clays. Peat actually represents an accumulation of disintegrated plant remains, preserved under condition of incomplete aeration and high water

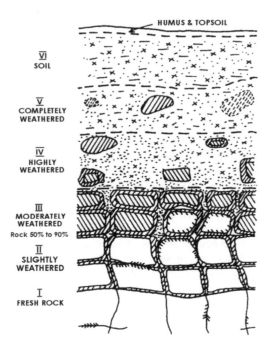

Figure 1.3.   Typical weathering profile in granitic rock soil profile (Little 1969).

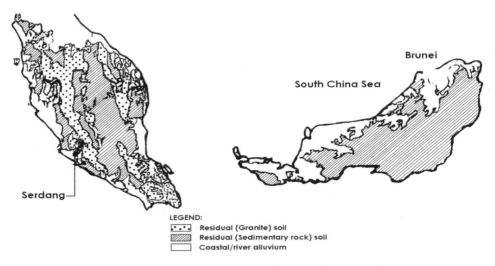

Figure 1.4.   Simplified soil map of Malaysia (Huat et al. 2002).

content. It accumulates wherever conditions are suitable, that is, in areas of excessive rainfall and where the ground is poorly drained, irrespective of latitude or altitude. Nonetheless, peat deposits tend to be most common in those regions with a comparatively cool wet climate. Physico-chemical and biochemical process cause this organic material to remain in a state of preservation over a long period of time. In other words, waterlogged poorly drained conditions, not only favor the growth of particular types of vegetation but also help preserve the plant remains.

Peat and organic soils represent the extreme form of soft soil. These soils are generally regarded as problematic soils as they are subject to instability such as localized sinking and slip failure,

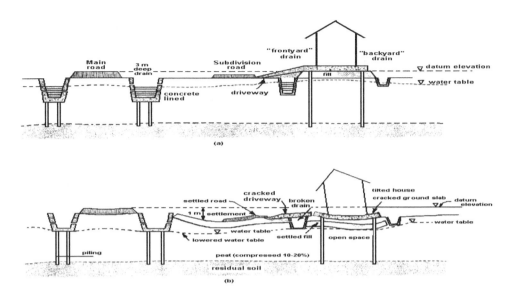

Figure 1.5. A cross section of a housing estate on peat (a) immediately after completion of construction, (b) several years after completion of construction (scale exaggerated) (Huat 2004).

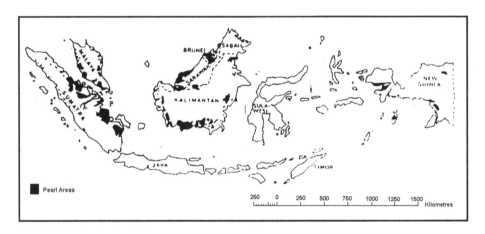

Figure 1.6. Tropical peat lands of Southeast Asia (Huat 2004).

and massive primary and long-term settlement when subjected to even moderate load increase. Buildings on peat are usually suspended on piles, but the ground around it may still settle, creating a scenario as depicted in Figure 1.5.

In addition, there is discomfort and difficulty of access to the sites, tremendous variability in material properties and difficulty in sampling. These materials may also change chemically and biologically with time. For example, further humification of the organic constituents would alter the soil mechanical properties such as compressibility, shear strength and hydraulic conductivity. Lowering of ground water may cause shrinking and oxidation of peat leading to humification with a consequent increase in permeability and compressibility. In short, these types of soils bring many problems to engineering, especially to the foundation.

Nearly two-thirds of the world's tropical peat lands of about 30 million hectares, are found in Southeast Asia (Figure 1.6). In terms of thickness, these deposits vary from just a few meters from the ground surface to tens of meters (Huat 2004).

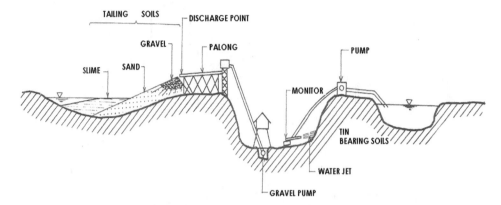

Figure 1.7.   Formation of tailing soils in gravel pump mining (Chan & Hong 1985).

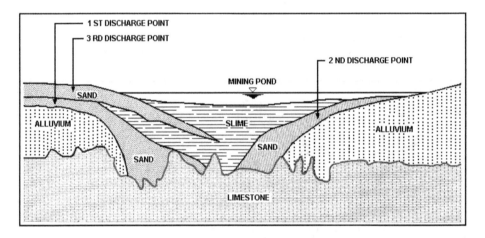

Figure 1.8.   Possible subsurface profile in ground subjected to gravel pump mining (Chan & Hong 1985).

The ex-mining land can also be categorized as soft soil. The mining industry in Malaysia (or specifically the tin mining industry) for example, can be traced back many years. As a result, the mining process has disturbed vast areas of natural ground. The mining activity leaves behind ponds, loose sandy soils, and slime deposits in the pond or on land. Slime is waste from mining and comprises very soft silty clay containing some fine sand. Figure 1.7 illustrates the process in the formation of tailing soils during a gravel pump mining. The tin-bearing soils are slurry that is then washed into a sump from where it is pumped to the top of the "palong". Here the heavy tin ore is separated from the rest of the soil slurry which now is referred to as the tailings, a waste product. This process produces two main types of soils: (i) loose to very loose sand and some gravel near the discharge point, and (ii) soft to extremely soft silty clay, often referred to as slime. To complicate matters, the discharge point is shifted now and then, and the end result may resemble the soil profile shown in Figure 1.8. Considering the nature of alluvial tin mining, it is clear the degree of disturbance is very severe resulting in drastic changes to particle size distribution, and hence changes in all the geotechnical properties of the soil. Note that mining by the dredging method produces a significantly different soil profile. Often the same ground may have been mined twice or possibly even three times using one or both methods of mining. Thus the tailing soil profile can be expected to be complicated and this is a characteristic feature of ex-mining land.

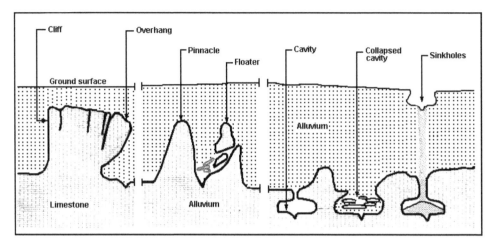

Figure 1.9. Karstic features of limestone bedrock (Chan & Hong 1985).

In Malaysia, the underlying rock of the tin mining area is usually limestone, of depths that vary from about 5 m to 50 m below the ground surface. Due to the weathering process, the limestone generally develops karsts features such as cliffs, pinnacles, solution channels, cavities, overhangs and sinkholes as shown in Figure 1.9. For such areas, it is not an easy task to plan the subsurface investigation and interpret the results. Apart from this, karstic features also pose considerable difficulties not only in the design but also in the construction of both deep and shallow foundations.

Sinkholes are a hazard to both shallow and deep foundations including piling. A sinkhole is a natural phenomenon in which the ground surface suddenly sinks in resulting in a large hole being formed. The presence of solution channels and underground cavities (or caverns), are sometimes favorable for seepage and underground flow of water which, by gradual removal of parts of the overlying soil, eventually form the substantial void (cavity) in the soil. This void may remain intact for a considerable period due to arching action until some disturbing agency causes the soil arch roof to collapse suddenly. The disturbing agency may be man-made, e.g. vibration from pile driving or lowering of ground water level; or it may be natural, e.g. earth tremors. The mechanism of the formation of a sinkhole has been described in detail by Jennings et al. (1964).

1.3 CLASSIFICATION OF FOUNDATION

Foundations are generally divided into two categories: shallow foundation and deep foundation (Figure 1.10). Shallow foundations are foundations built near or at ground (soil) or rock surface. This foundation is placed on a firm soil near the ground and beneath the lowest part of the superstructure. Examples of these foundations are pad footing and spread or raft footing. Deep foundation on the other hand is foundation that transmits structural loads to deeper soil or rock layer that is far from the surface. This foundation is the one, which is constructed on a soil that is not firm, and transmits the load of the structures considerably below the ground of the lowest part of the superstructures. Deep foundation can be divided into the following categories: pile foundation, drilled shaft or bored pile foundation, caisson, auger cast piles, pressure injected footing and micro piles.

Piles are normally columns made of concrete, wood, steel or plastic, as well as composites such as concrete-filled steel pipe or plastic-steel composite, driven into the ground. Drilled shafts (also commonly known as bored piles) are special piles made of cast in situ reinforced concrete inside a bored or drilled hole. Caissons are prefabricated cylinders or boxes that are sunk into the ground to some desired depth, and then filled with concrete. Auger piles are constructed by drilling slender cylindrical holes into the ground using a hollow stem auger, and then pumping grout through the

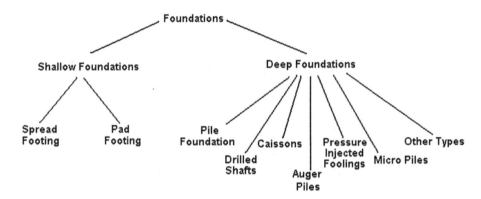

Figure 1.10.   Classification of foundation.

auger while it is slowly retracted. Pressure injected footings use cast in situ concrete that is rammed into the soil using a drop hammer. Micro piles or mini piles are small diameter piles formed by drilling and then the hole grouted with reinforcement of high tensile steel bars or pipes.

In Malaysia, the development of pile systems to suit the varied tropical ground conditions has come a long way (Ting 1998). The earliest commonly used piles to support up to medium rise structures in poor ground was the 300×300 mm and 375×375 mm square section reinforced concrete (RC) pile with grade 20 concrete, cast in situ as there was then no pile manufacturing facilities. For smaller structures and shallow depths (up to about 6 m) of poor ground with a high ground water table, *bakau* (mangrove tree, which is a type of timber pile) were used. Treated timber piles were later introduced. With the availability of modern pile manufacturing facilities, precast RC piles are now easily available. The production of driven pre-cast pre-stressed spun concrete piles, originally developed in Japan, is a fully developed industry in Malaysia, because of the ease in handling and its suitability in a wide range of formations found in Malaysia, ranging from soft clay/loose sands to the harder formations.

Drilled shafts, which are also known as drilled piers, or more commonly referred to as bored piles in Malaysia, have been found to be suitable in urban areas not only because of environment requirements but also to cater for the deep weathered formation. For heavier loads and larger diameters drilled shaft (up to 1.5 m), equipments have been developed to core suitable depths in underlying rock formations to form the pile socket (Ting et al. 2004). Hand dug caissons are sometimes used in steep terrain with limited space.

Micro piles (a.k.a. mini piles) of diameter 150–300 mm are used especially for underpinning of existing structures in distress, as well as support for structures (buildings and bridges) in difficult ground conditions, particularly in shallow karsts limestone formations, boulders in granite formations as well as soft ground over shallow hard formation (Ooi et al. 2004).

Patented pile systems such as the pressure injected footings (Franki pile) and auger piles, and steel H piles are also available.

In special circumstances such as footing for tank on soft alluvial ground, or deep basement with heavy loads over karstic limestone formation, raft and pile raft foundations, and barrette piles are used. To overcome the potential danger from slump zone at the contact between the pile toe and the limestone, compaction grouting is often carried out to infill the slump zone.

1.4   UNDERSTANDING THE BEHAVIOR OF FOUNDATION AND STRUCTURE

In foundation engineering, it is important that we draw on the experiences of past practices. As a matter of fact, many of the empirical methods that are used in foundation design are based on observations of the behavior of buildings and structures. Field observation is therefore important for a foundation engineer to understand the behavior of the ground, soil and structure. An understanding of the interaction between the ground and the structure requires precise measurements on the behavior of the ground below and around the foundation, and also the overall behavior of the superstructure. A good understanding of the behavior of the ground and its interaction with the foundation and structure will enable us to do a better design and thus reduce the overall construction cost.

The equipment required for field measurement have to be simple to operate, reliable, stable, cheap, easy to install and must be durable. Measurements that are normally made in observing the behavior of the foundation and structure are as follows:

(a) *Measurement of Vertical Movement*
Precise leveling is a simple technique for settlement or surface displacement. The reference station has to be stable and must be not in surface zone that can shrink or swell. Equipment such as the theodolite and levels can be used to measure vertical movement.

It would however be difficult for us to evaluate the characteristics of the subsoil just by measuring the surface settlement only. But the compression of the various subsoil layers can also be measured with equipment such as the deep settlement probes or extensometer. When combined with measurements of pore water pressure, the in situ characteristics of the soil at various depths can be determined.

(b) *Measurement of Lateral Movement*
The importance of lateral movement in observation of the foundation performance is often ignored. Lateral movement, however, is important especially around excavation, settling areas and foundation subjected to lateral loading. Extensometer can be used to measure this movement. As with the measurement of vertical movement, the value measured is more meaningful if lateral movements of the various depths are also measured in addition to the surface measurement. Surveying equipment such as the theodolite and EDM (electronic distance measurement) can be used for measuring surface lateral movement.

(c) *Load Measurement*
Load measurement is important in any studies on the soil-structure interaction. Although the main technique of the load measurement is properly understood, the un-conducive environment and the long time period require for observing a foundation usually made measurement like this difficult and expensive to make.

Equipments such as load cells can be used for this purpose. If a precise measurement of load is to be made on a foundation, then it is more appropriate to install the load cell in the structural element at the founding level.

(d) *Pressure Measurement*
Pressure measurement is actually one of the most difficult measurements to be made in particular for foundation. The presence of rigid boundary poses a special problem. Pressure cell is invented to bury in a fill, and is not ideally suitable for foundation. This is because the surface of the pressure cell needs to have similar characteristics to the surface of the materials surrounding it.

(e) *Movements in Building*
Another important aspect in observation of structure behavior is in limiting damage. This is best done using high quality photography; detail notes and sketches showing the pattern and width of the cracks. Changes in the width of the crack can also be observed using special equipment such as the Demec gauge.

1.5 DECIMATION AND PUBLICATION OF CASE STUDIES

As has been said above, a foundation engineer depends a lot on field observation to understand the behavior of the ground and structure. Therefore the publication of case studies in this subject can be said to be very important. It can be used as a guide to other engineers to design foundations and structures in other places of similar ground conditions.

But the usefulness of some of these case records can be limited due to absence of some vital information. Burland et al. (1977) in their Tokyo report listed down a list of important information that must be included in any case record.

a. Detailed soil profile together with its groundwater condition. Detailed descriptions of the soil including its consistency, structure, fabric, index values and others.
b. Results of probing tests (such as Macintosh probe) and other tests normally done in situ such as in situ index test.
c. Description of the sampling equipment and method used.
d. Results of laboratory tests, giving details of the test methods. Typical stress–strain curve, with "average" results given in the form of statistics.
e. Detailed results of in situ tests such as standard penetration test (SPT), cone resistance, field vane and plate loading.
f. Details of the foundation and structure including plan, cross-section, design load, actual loading and sequence of construction.
g. Details of instruments used, method of calibration and objective of evaluation.
h. Measurement of displacement, pressure and load, including closing error and differences in datum.
i. Detailed record on the performance of the structure and its finishing. This is best done using high-grade photography with good sketches.

The habit of observing and reporting the behavior of every building or structure has to be in-calculated in every foundation engineer as this is the only way to check from time to time the design assumptions made in order to elevate the level of confidence on the predictions made. Questions might be raised is as to what type of buildings need to be observed. Below is a list that can be used as guide.

1. Large structures with comprehensive site investigation.
2. Structures that are simple in shape or in plan or structures that are founded on uniform ground because this will simplify interpolation and comparison with the results of testing.
3. Structures that are founded on a soil stratum that is little known about its characteristics, or where there is little experience.
4. Structures where the load concentration and differential settlement may cause a major problem.
5. Structures with large undulating load.
6. Structures that might be badly affected by nearby proposed projects.
7. Structures where the movement is expected to continue leading to failure.

It is just as important to learn from the observation of any failures of the foundation, as the awareness would help to manage the same type of problems in some context. This work is best done by organizations such as local authorities, consulting firms and institutes of higher learning. Any organization that is actively compiling case records in their particular district will undoubtedly be making a major contribution to the profession. This is particularly so for the tropical region where knowledge and understanding of the soil-foundation behavior is lacking compared to those of the temperate zone soils. The need for knowledge on these soils however is great because of the extensive construction on these soils worldwide.

REFERENCES

Burland, J.B., Broms, B.B. & De Mello, V.F.B. 1977. Behavior of foundation and structures. *9th International Conference on Soil Mechanics and Foundation Engineering*. Tokyo (2): 495–546.
Chan, S.F. & Hong, L.P. 1985. Pile foundations in limestone areas in Malaysia. Foundation problems in limestone areas in Malaysia. The Institution of Engineers Malaysia. Malaysia: Kuala Lumpur.

Huat, B.B.K., Ali, F.H. & Maail, S. 2002. *Kejuruteraan geoteknik.* University Putra Malaysia Press. Malaysia: Serdang.

Huat, B.B.K. 2004. *Organic and peat soil engineering.* University Putra Malaysia Press. Malaysia: Serdang.

Jennings, J.E., Brink, A.B.A., Louw, A. & Gowan, G.D. 1964. Sinkholes and subsidence in the Transvaal dolomite of South Africa. *Proceedings of 6th International Conference on Soil Mechanics and Foundation Engineering,* Montreal. 1: 51–54.

Little, A.L. 1969. The engineering classification of residual tropical soils. *Proceedings of Specialty Session on the Engineering Properties of Lateritic Soil, 7th International Conference on Soil Mechanics and Foundation Engineering,* Mexico City. 1: 1–10.

Ooi, T.A., Tung, N.C. & Ng, S.L. 2004. The use of micro piles in Malaysia. *Proceedings Malaysian Geotechnical Conference. The Institution of Engineers Malaysia.* Kuala Lumpur. 323–334.

Ting, W.H. 1998. A survey of pile systems in Malaysia – past and present. Foundation course. The Institution of Engineers Malaysia. Malaysia: Sarawak.

Ting, W.H., Chan, S.F. & Ooi, T.A. 2004. Achievement of geotechnical engineering practice in Malaysia. Keynote lecture. *Proceeding of Malaysian Geotechnical Conference. The Institution of Engineers Malaysia.* Kuala Lumpur. 3–18.

CHAPTER 2

Geological Investigations for Foundations

Harwant Singh
Faculty of Resource Science and Technology, University Malaysia Sarawak, Malaysia

Husaini Omar & Bujang B.K. Huat
Department of Civil Engineering, University Putra Malaysia, Malaysia

2.1 INTRODUCTION

A sound understanding of the solid part of the earth on which structures are to be built is essential in civil engineering. An in-depth comprehension of the rocks and their characteristics is mandatory to build with confidence and have stable structures. No amount of well-designed superstructure will help if the foundations are on ground that has not been well understood and characteristics appreciated. A proper geological evaluation of a site is crucial. This requires gathering information on the geology of the ground as well as first-hand geological investigation.

2.2 OBJECTIVES OF GEOLOGICAL INVESTIGATIONS

The primary objective of geological investigations is to identify significant geological features and obtain the information required to design suitable foundations and build a safe structure. The geological investigation must reveal, at the least, the following:

– The rock types forming the anchor for the foundation
– The type, thickness and aerial extent of the soil layer
– The degree and extent of surface weathering
– The nature and depth of overburden to be removed for a foundation
– The engineering properties of the rock types for the foundation
– The geological structure of the subsurface foundation
– The history, type and extent of geologic deformation and structural defects in rocks due to tectonic forces
– The identification of present and potential geological hazards.

On obtaining the above information, geological investigations are able to assess the functional integrity of foundations.

Economically, according to Crawford (1962), two general reasons can be offered for investigating a building site from a foundation point of view are:

– The first, to permit design of the most economical, satisfactorily safe foundation for the proposed structure.
– The second, which also has significant economic overtones, is to provide sufficient reliable subsurface information to permit contractors to bid on the job without having to budget for uncertainties.

Ultimately, the owner benefits from lower bids, better job relations, fewer extras and the absence of lawsuits. Crawford (1962) states that the selection and design of foundations for a structure are as important as all other design considerations. It is just as important to know the properties of soils, as it is to know the properties of the materials in the superstructure. He further states that unlike the other building materials, the designer does not control the soil properties. The best that can be done is to assess them in situ.

2.3   SOURCES OF GEOLOGICAL INFORMATION

Maps

(a) *Geological maps*

Geological maps depict the spatial distribution of different types of geological information and understanding them is essential for understanding the geology of an area. The main types of geological maps are those that portray the stratigraphy or lithology of an area along with structural features. The former show the bedrock as units of geological time while the latter portray the bedrock differentiated into different lithologies.

Stratigraphy: These geological maps show the aerial distribution of the bedrock as units, called map units, as units of geological time. These units are chronostratigraphic units that have a direct equivalent in geological time units. Sedimentary rocks are shown as map units grouped according to their age of deposition regardless of lithology and igneous rocks according to their time of origin. The rocks, represented by their succession in time and age, are read in conjunction with the stratigraphic column shown on the map. The understanding of stratigraphy draws upon an understanding of some basic historical geology.

The Earth is now estimated to be 4.6 billion years old. The different rock types formed at different times have assigned ages of their time of formation from the present time based on the geological time units of the geological time scale. Stratigraphy determines the succession in time of rocks to establish the sequence in which the rocks were deposited or formed.

The duration of Earth history or geologic time is divided into intervals of time. This geological time scale with a hierarchy of time units is given in Table 2.1. An example of a time unit is the Cretaceous Period and one of its sub-divisions is the Campanian Age.

The age of map units according to the geological time units are designated as chronostratigraphic units or time-rock units. For example, rocks deposited during the Triassic Period are grouped into the Triassic System. The subdivisions of the time units and directly equivalent chronostratigraphic units are given in Table 2.2. The map units are read in conjunction with the stratigraphic column from the oldest units at the bottom to the youngest units at the top. The age of the rocks is determined by a number of stratigraphic techniques.

The information most often required for engineering purposes is the lithological and structural information. These stratigraphic rock units or map units normally do not have different constituent rock types individually demarcated but these may be found mentioned collectivity in the legend. This tends to drastically limit their usefulness for engineering purposes. However, in spite of this limitation there is one map unit referred to as the Quaternary that does assist directly in engineering site investigations of sedimentary rocks.

Lithology: These geological maps portray bedrock differentiated on the basis of their lithology. The bedrock is differentiated as lithological rock units that are separated by lines called contacts or geological boundaries. These rock units are referred to as lithostatigraphic units and there is a hierarchy of these rock units each demarcated by lithological criteria set for their composition. The understanding of lithology and sequence of deposition draws upon the understanding of petrology and sedimentology.

The fundamental lithostratigraphic rock unit is the Formation. A Formation is defined as a prevailingly body of rock identified by lithic characteristics that can be mapped in the field as a continuous unit. Formations can be, although they do not have to be, divided into members or grouped into groups as in Table 2.3.

Table 2.1.   Geological time scale showing time units (http://www.geosociety.org/science/timescale/timescl.htm).

1999 GEOLOGIC TIME SCALE

Table 2.2. Time units and chronostratigraphic units.

Time units		Chronostratigraphic units	
Era		Erathem	
Period	Late Middle Early	System	Upper Middle Lower
Epoch Age Time		Series Stage Substage	

Table 2.3. Lithostratigraphic units.

Lithostratigraphic unit	Description
Group	A collection of Formations
Formation	A physically continuous unit of strata that have a consistent lithic character that differs from adjacent units
Member	A subdivision within a Formation that is too thin to conveniently map but has distinctive lithologic characteristics.
Bed	Subdivision of a Formation or Members

A lithological map showing the spatial distribution and demarcation of the contacts between the different rock types largely fulfils engineering needs.

(b) *Engineering geology or geotechnical maps*
Thematic geological maps tailored for engineering purposes with pertinent information on lithological properties are essential. These geotechnical features that should be displayed are, for instance, general foundation suitability of different terrain for preliminary assessment on suitability for foundation work and relative compressibility/settlement of different terrain.

These maps are defined as a type of geological map that provides a generalized representation of all the components of a geological environment of significance in landuse planning, design, construction and maintenance as applied to civil engineering (The Unesco Press 1976). It is a synergetic product of two approaches of investigating the subsurface, namely, geological and geotechnical. Examples of such maps are the general foundation suitability map, and the relative compressibility/settlement map as shown in Figures 2.1 and 2.2 respectively.

The general foundation suitability map classifies the unconsolidated cover into classes of differing suitability for general foundation work. This classification allows a preliminary assessment of the requirements and constraints for general foundation work like the condition of ground surface, settlement and depth to foundation level forming the base for detailed site investigation.

In the example given in figures, foundation suitability is classified into four classes as follows:

1. *Very suitable*: These are areas underlain by in situ deposits (clay, sandy to clayey silt and clayey sand) formed from the weathering of igneous intrusive, sedimentary and metamorphic rocks. They have a shallow foundation level or rockline (1 to 10 m) with no danger of subsidence expected and a little or no fill required.
2. *Suitable*: These are areas underlain by shallow estuarine/deltaic deposits (clay, silt with lesser amounts of sand and gravel and variable organic matter) and riverine deposits (sand, silt and clay with minor to moderate plant remains) of less than 5 m sand. Shallow peat (less than 1 m) normally occurs as narrow belts. These areas have a moderately deep foundation level or rockline (10 to 15 m) with minor to moderate amounts of subsidence expected and fair amount of fill required.

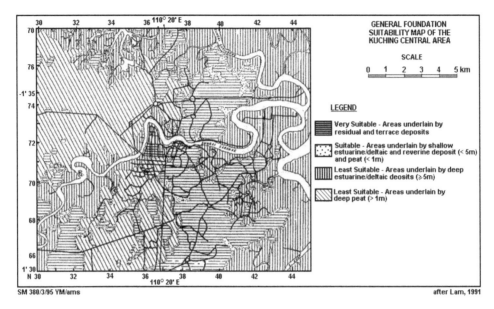

Figure 2.1.   Example of a general foundation suitability map for the Kuching area, Sarawak, Malaysia (after Lam 1991).

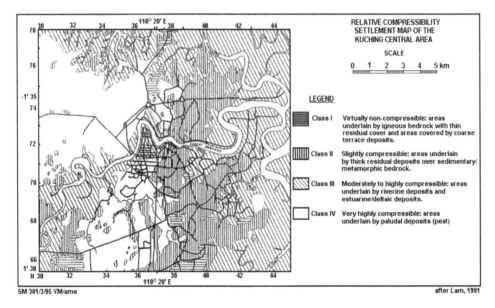

Figure 2.2.   Example of relative compressibility map for the Kuching area, Sarawak, Malaysia (after Lam 1991).

3. *Less suitable*: These are areas underlain by estuarine/deltaic deposits of more than 5 m thick. They are below flood/tidal level and have a deep foundation level or rockline (15 to 30 m) that require a moderate amount of fill.

4. *Least suitable*: These are areas underlain by deep peat (≥ 1 m). They are above flood level but are always water-logged with a very poor surface foundation (very soft and wet) and have a deep foundation level or rockline (12 to 20 m). Very severe subsidence is to be expected and a large amount of fill is required.

The relative compressibility/settlement map (Figure 2.2) refers to the susceptibility of earth materials to decrease in volume when subjected to load. High compressibility values are indicative of poor foundation stability and excessive ground settlement that determine the suitability for various land uses.

In the example given above, the map is divided into four relative compressibility classes as follows:

1. *Class I: Virtually non-compressible*: These are areas underlain by igneous bedrock with thin superficial cover and areas covered by coarse terrace deposits. The igneous rocks occur as small intrusive stocks forming low isolated hills and numerous dykes and sills. The terrace deposits consist of boulders, pebbles and coarse sand in a silty fine sand matrix.
2. *Class II: Slightly compressible*: These are areas characterized by in situ soils formed from the weathering of sedimentary and metamorphic rocks. Their thickness ranges from 3–12 m and are composed of clay, sandy clay or clayey sand. They are rather dense so only slight compaction is expected except in very clayey areas where the compressibility could be higher.
3. *Class III: Moderately to highly compressible*: These areas are characterized by riverine and estuarine/deltaic deposits which are composed of clay, silt and minor sand rich in organic matter that has an extremely high moisture-retention capability. This will result in high compressibility. As the marine clay of these estuarine/deltaic deposits is saturated and the permeability of clays is low, the consolidation settlement resulting from pore water being squeezed from the clay is generally very slow causing long continued settlement.
4. *Class IV: Very highly compressible*: These areas are characterized by paludal deposits or peat made up of slight to moderately decomposed plant remains and water. These deposits are very highly compressible due to their extremely high water content (80 to more than 90%) and organic content.

Reports

(a) *Geotechnical investigation reports*
These reports from nearby projects or previous projects on the current site can offer a lot of information especially data from tests performed previously.

(b) *Soil survey reports*
Although these reports are developed primarily for agricultural purposes, they contain maps with classifications and information on near-surface conditions from investigations that may be useful. In tropical areas, for example, these reports may be essential to obtain dimensions of soft soils such as peat.

Site investigations

The importance of a proper geological evaluation of a project site cannot be over-emphasized. The investigation of the site geology i.e. rock types and their extent, the weathering types and magnitude, etc. is a must. A detailed investigation involves surface mapping, drilling boreholes and trenches or using geophysical surveys. It is one of the best assurances against potential adverse consequences.

(a) *Site inspection*
A start should be made by undertaking an onsite inspection with the objective of getting a feel of the terrain and determining investigations needed to collect necessary data for evaluating the site conditions. The geologist and geotechnical engineer should jointly make detailed observations regarding the character of the subsurface. This then enables planning for sampling and other testing requirements for a full investigation.

(b) *Site investigations*
The objective of site investigations is to evaluate the subsurface conditions by obtaining samples and in situ testing of the ground. This provides primary data to base inferences and decisions for construction.

Subsurface investigations and sampling: The objective is to determine the ground conditions for geotechnical design, performance and detect potential difficulties and problems. There are various methods to obtain samples that include the digging of test pits, trenches and boreholes.

– Test pits are rectangular slots excavated in the ground with hydraulic backhoes for examination of shallow surficial materials. Their sides are systematically logged through soil and rock descriptions, photographs and sketches. Samples are taken and in situ measurements are made of properties like strength and density.
– Boreholes are vertical holes, drilled into the ground to obtain subsurface samples. Different techniques are used for soils and rocks.

Drilling and sampling in soils:
– Solid auger methods are used for shallow explorations of up to 3 m where auguring is done with a hand auger or a small drill rig. After being drilled a certain distance into the ground, the auger is then pulled up for the material on it to be examined and described. Samples are taken with thin-walled sampling tubes.
– Hollow stem auger method uses casing around the borehole sides to holdback the material with sampling done at the bottom of the casing. Greater depths are sampled using this method. Undisturbed samples are taken with thin-walled sampling tubes.

Drilling and sampling in rock: More force has to be applied to bore a hole in rocks as they are firmer and harder to penetrate. Some of the methods used are:

– Cable-tool method in which a heavy tool is dropped into the hole to break up the rock at its bottom. A drilling mud is circulated through the hole which then lifts chips of broken rock to the opening.
– Air rotary method is the same as the above method except that compressed air is injected into the hole to blow out the rock chips.
– Rock coring method whereby a hollow rock core with a diamond bit all around its circumference at the end drills through the rock. Water and drilling mud has to be circulated in the hole in order to cool the drill bit and remove rock chips clogging the hole. This rock core is pulled out after drilling a certain distance.

The undisturbed samples obtained are suitable for evaluating engineering properties like strength and compressibility while the disturbed samples are only suitable for soil identification and classification.

In situ tests: In situ testing measures the soil or rock properties in place. They are essential as this decreases disturbance of the material sampled due to the sampling process. Some of the in situ tests are as follows.

a. The Standard Penetration Test (SPT): This is a measure of resistance to penetration when a drive split spoon sampler is driven in with a hammer. The number of hammer blows to drive the split-spoon sampler is counted. The SPT values and example of correlations between two soils types are given in Table 2.4.
b. The Cone Penetration Test (CPT): This is the measurement of the resistance at tip and friction along sides when an instrumented cone is pushed into the ground.

Table 2.4.   SPT values and correlations (after Terzaghi & Peck 1967).

Sands		Cohesive soils	
Blows		*Blows*	
0–4	Very loose	0–1	Very soft
5–10	Loose	2–4	Soft
11–20	Form	5–8	Firm
21–30	Very firm	9–15	Stiff
31–50	Dense	16–30	Very stiff
>50	Very dense	>30	Hard

Note: A smaller version of the SPT equipment that is normally used for quick exploratory investigation is the Mackintosh or JKR probe.

c. Vane Shear Test: This test measures the shear resistance of soils. It consists of forcing a vane with four orthogonal blades into the soil and rotating it until the soil fails. The maximum torque value is measured.
d. Dilatometer Test: This test determines the in situ lateral stress in the subsurface. A dilatometer with a flexible membrane is pushed into the ground and is then inflated with compressed air from the surface. As the pressure is increased, the lateral earth pressure pushing against the membrane causes the disk to lift off from its starting position. The pressure that causes this movement is lateral stress.

2.4   SUBSURFACE CHARACTERISTICS

The Earth's crust constitutes the subsurface and is the domain of most engineering practices. The crust is formed from materials called rocks often covered with a veneer of unconsolidated material. This consequently results in two subsurface types.

Subsurface types

(a) *Bedrock*
The part of the crust constituted from consolidated rocks is referred to as the bedrock. These rocks are also referred to as intact rocks. The physical properties of fresh (unweathered) rock depend on their constituent minerals and texture and the presence of structural discontinuities like fractures, joints and bedding.

There are two types of intact rocks designated by the terms rock mass and rock material. The rock material refers to intact rocks that contain no joints. Rock mass refers to the bedrock that consists of intact rock with joints.

Each of these main rock types are tremendously varied and classified on the basis of composition and structure in addition to the mode of origin. Based on their mode of origin, rocks are broadly classified as igneous, sedimentary and metamorphic rocks.

Igneous rocks form by the cooling and crystallization of hot, molten magma or lava from within the Earth solidifying at or near the surface. There are a wide variety of igneous rocks and they may be classified according to a number of criteria. Those formed at depth are termed intrusive or plutonic and those formed on the surface are called extrusive or volcanic rocks as shown in Table 2.5.

Extrusive igneous rocks that cool rapidly are texturally very fine-grained whereas slowly cooled intrusive igneous rocks are coarse-grained. A glassy texture results when the rock cools too quickly for crystals of minerals to form.

Sedimentary rocks are those formed by compaction (lithification) of deposited weathered material on the surface of the Earth. They, therefore, form from particles or sediments of pre-existing or

Table 2.5.   Igneous rocks.

Plutonic rock	Equivalent volcanic rock
Granite	Rhyolite
Granodiorite	Dacite
Tonalite	Quartz andesite
Syenite	Trachyte
Monzonite	Latite
Diorite	Andesite
Gabbro	Basalt

Table 2.6.   Clastic sedimentary rocks.

Composition		Grain size	Rock name
Mainly quartz, feldspar, clay minerals and rock fragments		Gravel (Over 2 mm)	Conglomerate Breccia
		Sand (0.0625–2.00 or 1/16–2 mm)	Sandstone
	Mud	Silt (0.0039–0.0625 or 1/256–1/16 mm)	Siltstone Shale
		Clay (Lesser than 0.0039 or 1/256 mm)	Claystone Shale

Table 2.7.   Non-clastic sedimentary rocks.

Mineral composition	Rock name
Calcium Carbonate, $CaCO_3$	Limestone
Dolomite, $CaMg(CO_3)_2$	Dolostone
Quartz, SiO_2	Chert
Halite, NaCl	Rock salt
Gypsum, $CaSO_4 \, 2H_2O$	Rock gypsum
Plant fragments	Peat and coal

parent rock – igneous, metamorphic or sedimentary rocks themselves. The parent rocks undergo chemical and/or physical weathering into sediment, and are transported by natural agents and deposited in sedimentary basins. Their mode of origin enables their subdivision into the three following types of sedimentary rocks:

• Clastic: formed from cemented sediment grain
• Chemical: formed from precipitation
• Biogenic: formed from organic matter.

The latter two are usually referred to as non-clastics. The primary minerals and textures of sedimentary rocks undergo modification or changes termed diagenesis, as the sediments are buried deeper due increasing temperature and pressure causing them to be lithified. Examples of sedimentary rocks are given in Tables 2.6 and 2.7.

Metamorphic rocks are formed due to pre-existing or parent rocks (igneous, sedimentary or metamorphic) and undergo solid state re-crystallization due to changes in pressure, temperature or chemistry (such as fluid addition). The type of metamorphic rock formed is determined by metamorphic conditions (specific combination of pressure and temperature) and the parent rock. Some examples of metamorphic and their parent rocks are given in Table 2.8.

Table 2.8. Metamorphic rocks.

Metamorphic rock	Precursor rock
Quartzite	Quartz sandstone
Greenstone	
Amphibolite	Basalt or Gabbro
Hornfels	Siltstone
Marble	
Serpentinite	Peridotite/Ultramafic igneous rocks
Soapstone	
Graphite	Anthracite coal

Table 2.9. Foliated metamorphic rocks.

Rock names	Diagnostic features
Slate	Well developed slaty cleavage
Phyllite	Well developed phyllitic texture; silky, shiny luster
Schist	Schistosity. Different types of schist apportioned on the basis of mineral content
Gneiss	Gneissic banding i.e. well-developed color banding due to alternating layers of different minerals

Table 2.10. BS 5930: 1981 igneous rock classification.

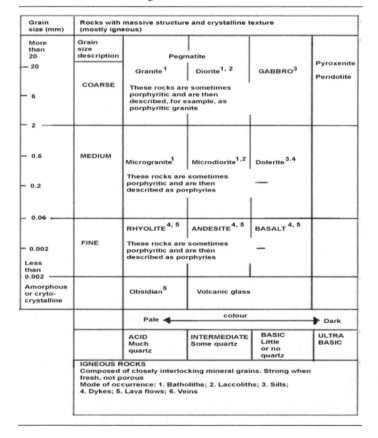

Texturally, some metamorphic rocks develop a fabric of parallel alignment or orientation of minerals or mineral banding. These are called foliations. Hence metamorphic rocks are divided into two main groups termed foliated or non-foliated on the basis of the presence or absence of this foliation.

These two groups are then sub-divided with the foliated rocks being classified based on different types of foliation and the non-foliated rocks termed on the basis of their composition. The different types of foliated metamorphic rocks are as found in Table 2.9.

There is great variation of physical properties between rock types with igneous rocks being least variable and sedimentary rocks being highly variable.

The physical properties of rocks are also not uniform over the entire rock mass. For example strength is not applicable to the entire rock mass as to a large extent it is controlled by the nature of the discontinuities or planes of weakness present like bedding planes and sedimentary structures (types of beds) in sedimentary rocks, foliation in metamorphic rocks, flow contacts, intrusive contacts, dikes and sills in igneous rocks and structural features like faults, joints and fissures.

Engineering requirements require rock classifications to apportion them on their properties pertinent to engineering needs. Some engineering classifications e.g. those of the International Association of Engineering Geology (IAEG 1981) and the BS 5930: 1981 classifications have been formulated to meet this necessity. The BS 5930: 1981 classifications for the three different rock types are as shown in Tables 2.10, 2.11 and 2.12.

With regard to the engineering properties of rocks, the essential quantifiable properties required to be determined include their strength, density, hardness or compressibility and the occurrence and incidence of discontinuities. These properties determine their behavior especially response to stresses.

The mechanical strength and hardness parameters are dictated by the physical characteristics of the various constituents that make up the rock. Some of these are density, bonding and cementation. Generally, the rocks with the highest mechanical strength are igneous or metamorphic rocks while sedimentary rocks tend to range from hard to the very soft.

As the strength of a rock is directly proportional to density, the densest rocks are normally the strongest. Density is determined by its crystallanity, compaction or cementation.

Rock strength is measured by its response to compressive stress that in turn gives compressive strength. The compressive strength of a rock is measured by the compressive stress that a specimen is able to withstand before breaking. The compressive strength measured is the uniaxial compressive strength where the load is applied vertically to a specimen unconfined on its sides as shown in Figure 2.3. It is also called the unconfined compressive strength. The units used to measure the compressive strength are N/m^2 (SI system) or PSI (British system).

The elastic moduli measure the compressibility of a rock i.e. the contraction it undergoes when subjected to a load. This is given by the Young's modulus, a material constant, which describes the stiffness or how easily the material deforms. It is the proportionality constant, E, in Hooke's Law: $\sigma = Ee$, that relates stress and strain and can also be expressed as follows.

$$E = \frac{stress}{strain} = \frac{\sigma}{e} \tag{2.1}$$

When deformation or strain is recoverable, then behavior is said to be elastic.

Engineering requirements require rock classifications to recognize and apportion them on quantifiable physical properties. Intact rock has been classified on mechanical properties such as strength and compressibility thus giving an indication of its engineering utility. One such classification of intact rocks is as shown in Figures 2.4, 2.5 and 2.6.

Various efforts are continuing for a standardized functional geotechnical scheme and a good account of this is available in Johnson & De Graff (1988).

(b) *Bedrock cover*
In the case of a tropical region such as Malaysia, the bedrock cover can be broadly divided into residual soil cover and transported soil cover.

Table 2.11.   BS 5930: 1981 sedimentary rock classification.

Grain size (mm)	Bedded rocks (mostly sedimentary)				
More than 20	Grain size description		At least 50% of grains are of carbonate	At least 50% of grains are of fine-grained volcanic rock	
20— 6— 2—	RUDACEOUS	CONGLOMERATE Rounded boulders, cobbles and gravel cemented in a finer matrix Breccia Irregular rock fragments in finer matrix	LIMESTONE and DOLOMITE (undifferentiated) — Calcirudite	- Fragments of volcanic ejecta in a liner matrix - Rounded grains AGGLOMERATE - Angular grains VOLCANIC BRECCIA	SALINE ROCKS Halite Anhydrite
0.6— 0.2— 0.06—	ARENACEOUS / Coarse / Medium / Fine	SANDSTONE Angular or rounded grains, commonly cemented by clay, calcitic or iron minerals Quartzite Quartz grains and siliceous cement Arkose Many feldspar grains Greywacke Many rock chips	Calcarenite	Cemented volcanic ash TUFF	Gypsum
0.002— Less than 0.002	ARGILLACEOUS	MUDSTONE / SILTSTONE Mostly silt SHALE Fissile / CLAYSTONE Mostly clay	Calcareous mudstone / Calcisiltite / Calcilutite / CHALK	Fine-grained TUFF Very fine-grained TUFF	
Amorphous or crypto-crystalline		Flint: occurs as bands of nodules in the Chalk Chert: occurs as nodules and beds in limestone and calcareous sandstone			COAL LIGNITE
		Granular cemented except amorphous sandstone			
		SILICEOUS	CALCAREOUS	SILICEOUS	CARBON-ACEOUS
		SEDIMENTARY ROCKS Granular cemented rocks vary greatly in strength, some sandstones and stronger than many igneous rocks. Bedding may not show in hand specimens and is best seen in outcrop. Only sedimentary rocks, and some metamorphic rocks derived from them contain fossils. Calcareous rocks contain calcite (calcium carbonate) which effervesces with dilute hydrochloric acid.			

Residual soils are soils that have been developed by in situ weathering of parent rocks. In the tropics, heavy rains and high temperature has lead to intensive chemical weathering leading to formation of tropical soils of several tens of meters deep. The profiles of these soils typically vary in a gradational manner with depth as shown in Figures 2.7 and 2.8. Detailed description of the origin, formation and occurrence of tropical residual soils can be found in Singh & Huat (2004).

Transported soils are soils that has been transported and deposited by agents such as water, wind and gravity. The profiles of transported soils have a complex geometry and a different lithology. Therefore, average quantitative values might not reflect weaker sections.

(c) *The Quaternary map unit*

The Quaternary map unit is a chronostratigraphic unit denoting an interval of time from the present to about 2.0 million years ago. Some geological maps depict this period as two chronostratigraphic units called the Pilestocene and Holocene. The quarternary cover, therefore, refers to deposits of a certain time period and its engineering significance is that the short geological time interval has not enabled the deposited material to be sufficiently indurated into rocks. It is present as a mantle of unconsolidated to semi-consolidated gravel, sand, mud and clay (Stauffer 1973) and can be observed directly on geological maps and found in literature for assessment for engineering purposes.

Table 2.12.   BS 5930: 1981 metamorphic rock classification.

Grain size, mm	Obviously foliated rocks (mostly metamorphic)		Rocks with massive & crystalline textures.
More than 20	Grain size description		
20 — 6 —	COARSE	GNEISS Well developed but often widely spaced foliation sometimes with schistose bands Migmatite Irregularly foliated: mixed schists and gneisses	MARBLE QUARTZITE Granulite HORNFELS Amphibolite Serpentine
2 — 0.6 — 0.2 —	MEDIUM	SCHIST Well developed undulose foliation; generally much mica	
0.06 — 0.002 — Less than 0.002	FINE	PHYLLITE Slightly undulose foliation; sometimes 'spotted' SLATE Well developed plane cleavage (foliation)	
Amorphous or crypto-crytalline		Mylonite Found in fault zones, mainly in igneous and metamorphic areas	
	CRYSTALLINE		
	SILICEOUS		mainly SILICEOUS
	METAMORPHIC ROCKS Most metamorphic rocks are distinguished by foliation which may impart fissility. Foliation in gneisses is best observed in outcrop. Non-foliated metamorphics are difficult to recognize except by association. Any rock maked by contact metamorphism is described as a 'norniels' and is generally somewhat stronger than the patent rock Most fresh metamorphic rocks are strong although perhaps fissile		

As this cover forms coastal lowlands and river valleys as a consequence of fluvial processes, it occupies areas of urbanization and infra-structural development. Hence, due to it being the foci of economic activity, it is a very important area for constructional activities. Major subsurface problems arising from soft soils are also to be found in this map unit.

According to Stauffer (1973) the Quaternary cover is found on coastal lowlands and constitutes the floors of some inland valleys. It also finds localized expression on terraces or as remnants of erosion deposits at higher levels.

Structure

(a) *Geological structures*
The Earth's crust is made up of subsurface geometrical bodies or features constituted of rocks that are termed as structures that arise as a result of the operation of forces in the crust called tectonic or

Figure 2.3. Compressive stress.

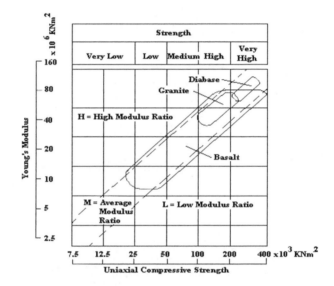

Figure 2.4. Classification for intact igneous rock (modified from West 1995).

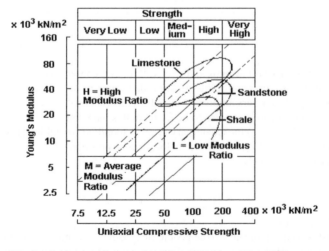

Figure 2.5. Classification for intact sedimentary rock (modified from West 1995).

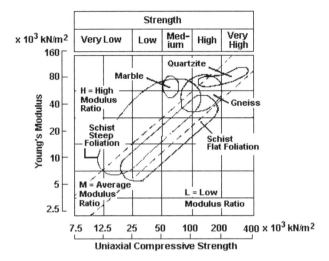

Figure 2.6.   Classification for intact metamorphic rock (modified from West 1995).

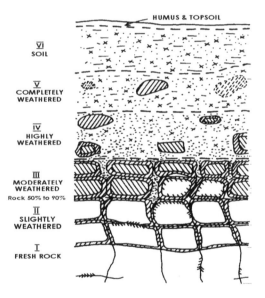

Figure 2.7.   Typical weathering profile in granitic rock (Little 1969).

crustal forces. These structural features or, just simply, structures have spatial forms and alignments as an outcome of emplacement and deformation of rock bodies that extent from microscopic to kilometers in size and extent. In addition depositional structures due to surface processes form on the crusts that are later in turn subjected to tectonic forces.

Bedrock cover structures: These are mainly depositional sedimentary structures formed during the deposition of transported material in depositional environments over pre-existing bedrock. These constitute three-dimensional geometrical bodies whose exposed surfaces exhibit characteristic geomorphological forms or features peculiar to that environment. Vertical cross sections reveal the sedimentary structures of these depositional environments normally encountered in site investigations. Some vertical profiles are given in Figures 2.9 and 2.10.

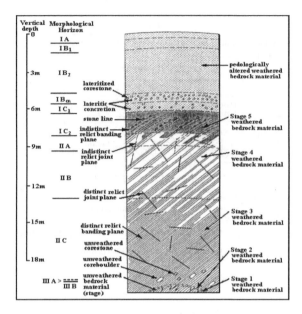

Figure 2.8.   Schematic sketch of different stages of the weathering profile over amphibole schist bedrock (Raj 1994).

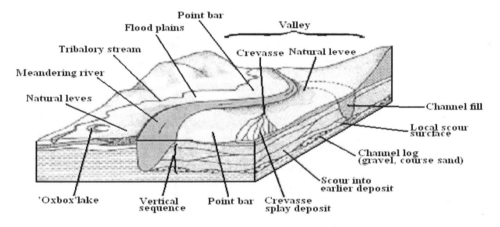

Figure 2.9.   Depositional structures in a fluvial environment. (*http://www.carleton.ca/~tpatters/teaching/ intro/depoenvironments/depoenvironments7.html*)

The result of continued deposition over a very prolonged period of time and consequent burial gives rise to form sedimentary rocks.

Emplacement structures: These structures are formed by the intrusion of magma into pre-existing, also called, country rock and subsequent cooling to igneous rocks of various shapes. A tabular intrusion forms a horizontal feature called a sill, one cutting across the country rock is termed a dyke and a large, discordant intrusive body is called a batholith or pluton. Lava i.e. magma escaping onto the surface forms volcanic structures or features. The above structures are shown in Figure 2.11.

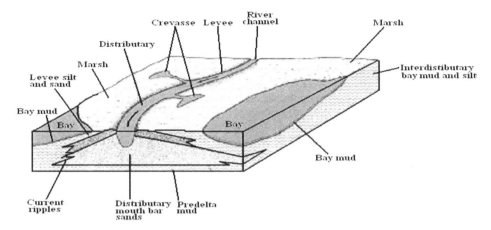

Figure 2.10.   Depositional structures in a delta.   (*http://www.carleton.ca/~tpatters/teaching/intro/depoenvironments/depoenvironments9.html*)

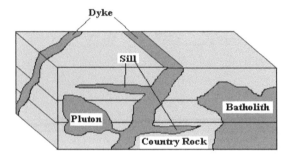

Figure 2.11.   Plutonic emplacement structures.

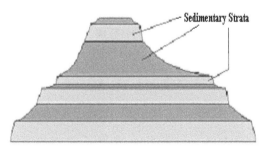

Figure 2.12.   Horizontal sedimentary strata.

Deformational structures: Deposition of sediments gradually gives rise to almost horizontal layers. These upon lithification form various strata of sedimentary rocks depending on the nature of dominant sediment as shown in Figure 2.12.

Stresses, due to tectonic activity, within the Earth's crust create subject rocks to bend, twist or fracture and consequently deform giving rise to rock bodies undergoing transformation in position, alignment, orientation and form. A working knowledge of subsurface rock structures is essential for engineering practice.

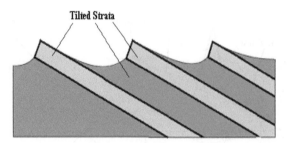

Figure 2.13.   Tilted strata.

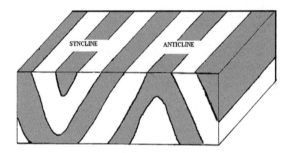

Figure 2.14.   Folds.

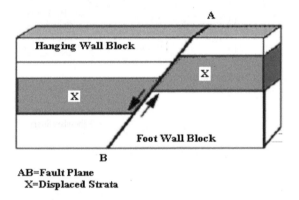

AB=Fault Plane
X=Displaced Strata

Figure 2.15.   Fault.

Tilted strata: This is a simple case of the strata tilted by tectonic forces resulting in their inclination to the horizontal as shown in Figure 2.13.

Folds: These are flexures formed when horizontal strata to fold upon compression as shown in Figure 2.14. The rising arches of the folded strata are called anticlines and depressed troughs are called synclines. On a large scale they give rise to domes and basins.

Faults: These are breaks or fractures in rocks as a result of brittle deformation as seen in Figure 2.15. The rocks show displacement on either side of the rupture or fault planes.

The rock strata or material above the fault plane is called the hanging wall block while that below is called the footwall block. The classification of faults on the basis of direction in which the blocks

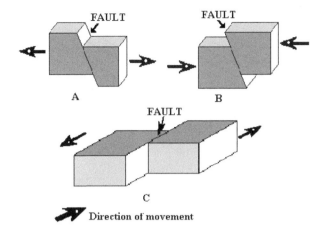

A: Normal fault; B: Reverse fault; C: Strike-slip fault

Figure 2.16.   Types of faults.

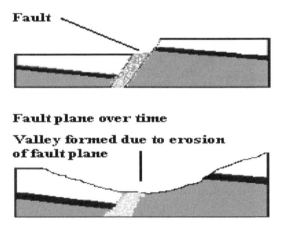

Figure 2.17.   Surface expression of a fault over time.

with respect to the fault plane enables the identification of three types of faults as described below and shown in Figure 2.16.

– *Normal Faults*: Normal faults form when the hanging wall moves downwards relative to the footwall along the fault plane.
– *Reverse Faults*: Reverse faults form when the footwall moves down relative to the hanging wall along the fault plane.
– *Strike-Slip Faults*: Normal faults form when the movement is horizontal along a generally vertical fault plane due to shear stress.

As the rocks forming the fault planes undergo erosion over time their resultant surface expression may be as given in Figure 2.17.

Joints: Fracture surfaces or breaks in a rock mass with no displacements along them are termed as joints are shown in Figure 2.18. They can range from microscopic fractures to kilometers in length

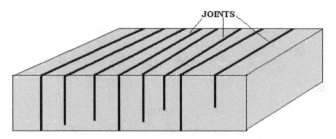

Figure 2.18.   Joints.

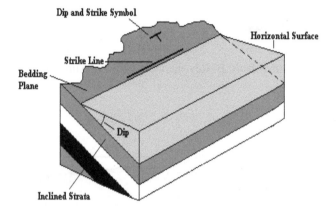

Figure 2.19.   Measurements of attitude of a fault.

with spacing ranging from centimeters to several meters. They may also occur as sets oriented in different orientations. In engineering terminology these are termed as discontinuities and rocks bodies with discontinuities are called rock mass while those without are called intact rock.

(b) *Geological structural information in geological maps*
The structural features are portrayed on geological maps through their expression on the surface. This information enables the configuration of the third dimension of subsurface to be visualized.

This information is gathered through fieldwork by locating the structural features and determining their orientation wherever there is a suitable structural exposure. For example, the attitude of bedding planes i.e. contacts between the different sedimentary strata provide the spatial orientation of these layers or offsets of strata indicate faults and their orientation. The attitude of bedding planes showing their spatial orientation including the direction and amount in which they extend under the subsurface is given by their dip and strike. The strike of the bed or strata is the direction of the line formed by the intersection of a horizontal plane (or surface) cutting its surface. The dip is the angle between the horizontal plane and the bedding (or planar) surface being measured perpendicular to the strike. These two measurements give the spatial orientation. This is shown in Figure 2.19.

Lines and symbols depict the structural information on the geological map. Lines mark features like the contacts between different map units or rock units and faults. The attitude of bedding planes showing their spatial orientation i.e. the direction and amount in which they extend under the subsurface is given by the dip and strike symbol along with the actual values. Some examples of map symbols for geological structures are given in Figure 2.20.

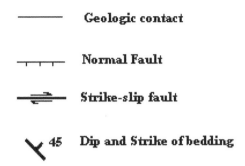

Geologic contact

Normal Fault

Strike-slip fault

45   Dip and Strike of bedding

Figure 2.20.   Examples of map symbols for geological structures.

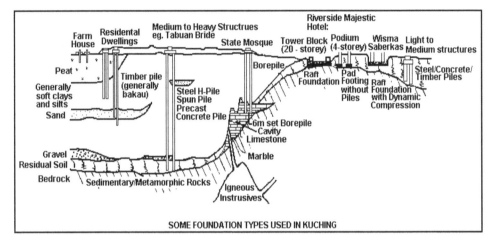

Figure 2.21.   Example of the different foundations and relationships to the rockline (Yogeswaran 1995).

Rockline

Knowledge of the depth to the rockline or rockhead is an important for the foundation works as it a factor in determining the condition of ground surface. Engineering works experience problems in foundation when structures are sited on deep peat or soft estuarine/deltaic deposits as settlement has caused cracking and even sinking of these structures. The differing bedrock depths are a component of the expected foundation costs for engineering works and also dictate the type of foundations to be used. The different types of foundations and their relationship to the rockline are shown in Figure 2.21.

(a) *Rockline or depth to bedrock*
The unconsolidated soils mantle on the bedrock has increased over time especially in tropical latitutes. As a result there may be considerable depth to the rockline or rockhead and the depth to the bedrock could also vary considerably. Precautions have also to be taken, as the degree of weathering may not be consistent throughout a construction site. An example is the different rockline depths may be seen from a hypothetical cross section in Figure 2.21.

(b) *Morphology of bedrock surface or rockline*
The buried bedrock or rockhead possesses a surface morphology due to differential weathering due to lithological variances and is not uniform. It is important for this to be mapped in sufficient

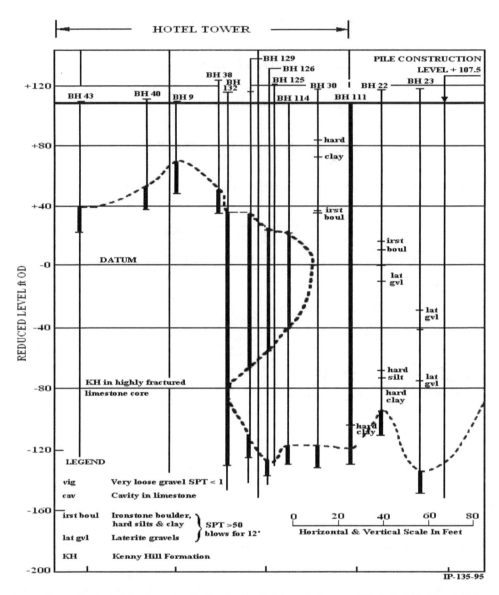

Figure 2.22.   Geological Section under the Pan Pacific Hotel, Kuala Lumpur (Mitchell 1985, Chow 1995).

detail. An example of this problem was encountered during the foundation investigations for the 30-storey Pan Pacific Hotel building in Kuala Lumpur, Malaysia (Mitchell 1985) where the limestone bedrock was found to have had a 15-metre overhang as seen in Figure 2.22.

Another example of instability in limestone bedrock is the presence of cavities. The relatively faster dissolution of limestone which gives rise to karst topography creating subsurface cavities that caves in to form sinkholes, as shown in Figure 2.23.

(c) *Lithological contacts*
This buried bedrock or rock head is also very liable to undergo a change in lithology at the contacts between the different rock types that again changes the characteristics of the subsurface. The

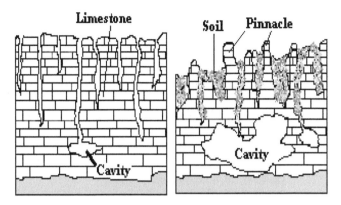

Figure 2.23.   Development of karst features and solution cavities.

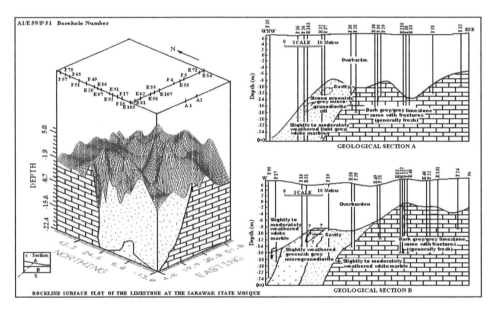

Figure 2.24.   Surface plot of the limestone and the geologic cross-sections of the North Kuching area in Sarawak, Malaysia (Yogeswaran 1995).

Figure 2.24 below is an example from a construction site where the rock type changes from igneous through metamorphic to sedimentary.

2.5   GEOLOGICAL HAZARDS

Geological hazards are naturally occurring events that pose or are a potential threat. Volcanic eruptions, earthquakes, landslides and flood cause serious damage to engineering structures. The assessment for the potentiality of a hazard along with the feasibility studies is essential for engineering projects. This mitigates the effect of future eventualities by the adaptation of suitable designs and rehabilitation with may involve high costs.

Table 2.13.   Mercalli scale (http://www.free-definition.com/Mercalli-Intensity-Scale.html).

Mercalli intensity	Characteristic effects	Approximate Richter magnitude
I – Instrumental	Not felt	1
II – Just perceptible	Felt only by a few persons, especially on the upper floors of tall buildings	1.5
III – Slight	Felt by people lying down, seated on a hard surface, or on the upper floors of tall buildings	2
IV – Perceptible	Felt indoors by many, few outside	3
V – Rather strong	Generally felt by everyone, sleeping people may be awakened	4
VI – Strong	Trees sway, chandeliers swing, bells ring, some damage from falling objects	5
VII – Very strong	General alarm, walls and plaster crack	5.5
VIII – Destructive	Felt in moving vehicles, chimneys collapse, poorly constructed buildings are seriously damaged	6
IX – Ruinous	Some houses collapse, pipes break	6.5
X – Disastrous	Obvious ground cracks, railroad tracks bent, some landslides on steep hillsides	7
XI – Very disastrous	Few buildings survive, bridges damaged or destroyed, all services interrupted (electrical, water, sewage, railroad, etc.), severe landslides	7.5
XII – Catastrophic	Total destruction, objects thrown into the air, river courses and topography altered	8

Seismicity

When rocks are stressed beyond their elastic limits they undergo a brittle fracture beneath the surface of the Earth. The release of energy gives rise to earthquakes. The point at which the rupture occurs is called the earthquake focus or hypocenter. The geographic point on the Earth's surface directly above the focus is called the epicenter.

The magnitude of an earthquake is measured according to the Richter Scale which is based on the amplitude of the seismic waves compensated for the loss of energy according to the distance that they traveled while the intensity is measured on the Mercalli Scale (Table 2.13) which is an arbitrary scale based on human observation and degree of damage. The Richter Scale is a logarithmic scale (base 10). Two earthquakes differing on the scale by 1 Richter unit differ in size by a factor of 10. An earthquake of 4 on the Richter Scale is 10 times greater than the one of 3 on the scale. So the one of 5 is 100 times greater than one of 3.

Earthquakes cause ground movement that causes destruction of rigid structures and liquefaction i.e. the sudden loss of strength of water saturated sediment. The ground movements also trigger mass movements like mudflows and landslides.

Malaysia is not located on any seismic belt but the tremor from seismic activity in neighboring Indonesia is often felt, including the massive earthquake and Tsunami that hit north of Acheh, Indonesia on December 26, 2004. Figure 2.25 shows the earthquake epicenters of the South East Asian region.

Slope stability and mass movement

A slope is an exposed ground surface that is positioned at an angle to the horizontal. In such a condition gravity will tend to drag material from the surface e.g. soil downward. If this happens

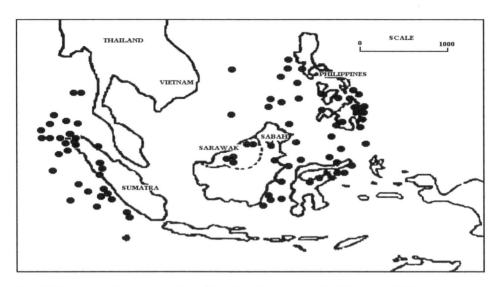

Figure 2.25.   Earthquake epicenters in the South East Asian region (after Yogeswaran 1995).

then the infraction of the state of the slope surface is referred to as a slope failure causing an infringement of the stability of the slope. Slope stability is very important for the utilization of the ground for construction. The stability analysis of a slope is very essential.

Although the controlling force of slope failure is gravity, many factors contribute to the instability of slopes. Amongst the main factors are the nature of the bedrock and the soil cover, the inclination and geometry of the slope and role of ground water. All these influence the driving forces and the resisting forces against mass wasting. For example, the driving force increases with the increase of water in the soil.

The down slope movement of rock and regolith due to gravity as a result of slope failure is termed as mass wasting. This mass wasting is occurs by the operation of mass wasting processes. The process of mass wasting is commonly referred to as landslides. However, there are various processes by which mass wasting occurs but as these processes generally grade into one another it is to some extent difficult to strictly separate these different processes.

There are two broad causes of mass wasting. One is a sudden failure of the slope sending loosened debris sliding, rolling, falling or slumping down hill and the other is a process called sediment flow where the debris flows down the slope mixed with water or air. Varnes (1978) further classified landslides by the type of movement and type of material involved as shown in Table 2.14.

The types of movement are slides, flows, lateral spreads, falls and topples which involves the following mode of transfer of materials.

– *Slides*. This movement involves the downward displacement of rock or soil through sliding along one or more failure surfaces.
– *Flows*. Flows consist of the movement of a slurry of loose rocks and soil down slope in the manner analogous to a viscous fluid. They are distinguished from slides as they have a high water content and are thoroughly deformed internally during movement (Kehew 1995).
– *Falls and Topples*. The incidence of masses of rock or other material detaching from a steep slope or cliff and descending by free fall, rolling or bouncing is referred to as a fall. When rocks or other material rotate forwards and downwards about a pivot point on a hillslope it is referred to as a topple. Topples generally develop in rock slopes that are divided into blocks by vertical fractures or joints oriented parallel to the slope face (Kehew 1995).

The types of material include bedrock and soil (including artificial fill). Soils are described as material that is either predominantly coarse (debris) or predominantly fine (earth).

Table 2.14.   Classification of landslides (Varnes 1978).

Type of movement			Bedrock	Engineering soils Predominantly coarse material	Predominantly fine material
				Type of material	
Falls			Rock fall	Debris fall	Earth fall
Topples			Rock topple	Debris topple	Earth topple
Slides	Rotational	Few units	Rock slump	Debris slump	Earth slump
	Translational	Many units	Rock slide	Debris slide	Earth slide
Lateral spreads			Rock spread	Debris spread	Earth spread
Flows			Rock flow (deep creep)	Debris flow (soil creep)	Earth flow (soil creep)
Complex movements			Combination of two or more of the principal types of movements		

Other hazards

(a) *Sinkholes*
Sinkholes are small, steep-sided depressions caused by the sudden collapsing of the surface into a cavern. As mentioned before, they are commonly found where the rock below the land surface is naturally be dissolved by ground water like limestone.

(b) *Flooding*
There are several different types of floods: flash, riverine and coastal. Due to the rainfall regime in tropical regions like Malaysia the riverine and low-lying areas are subjected to flooding. Flooding causes damage to infrastructure facilities and buildings. As a consequence of river channel dynamics riverbank erosion is also widespread and riverbed scouring is an inevitable part of the hydraulic process.

(c) *Sea level rise*
Submergence as a result of sea level rise is also a hazard for coastal areas. The determination of the general ground level of recent coastal deltaic and estuarine alluvia deposits is essential.

REFERENCES

British Standard Institution, BS 5930: 1981. Code of practice for site investigation, British Standards, London.
Chow, W.S. 1995. Geological inputs in civil works – Problems and issues. *Proceedings of the Dialogue Session on Geological and Geotechnical Considerations in Civil Works*. Geological Survey of Malaysia, Kuala Lumpur. 1–26.
Crawford, C.B. 1962. Engineering site investigations, CBD-29, Canadian Building Digest, 01-05-1962. Institute for Research in Construction, National Research Council of Canada.
IAEG, 1981. Rock and soil description and classification for engineering geological mapping. Int. Assoc. Eng. Geol. Bulletin. 24: 235–274.
Johnson, R.B. & De Graff, J.V. 1988. *Principles of engineering geology*. New York: John Wiley.
Kehew, A.E. 1995. *Geology for engineers and environmental scientists*. Second Edition, Englewood Cliffs, New Jersey: Prentice-Hall.
Lam, S.K. 1991. Quaternary geology of the Kuching Area, Sarawak, Map Report 9. Geological Survey of Malaysia.

Little, A.L. 1969. The engineering classification of residual tropical soils. *Proceedings of Specialty Session on the Engineering Properties of Lateritic Soil. 7th International Conference on Soil Mechanics and Foundation Engineering*, Mexico City 1: 1–10.

Mitchell, J.M. 1985. Foundation for the Pan Pacific Hotel in pinnacled and carvernous limestone. *Proceedings of 8th South East Asian Geotechnical Conference*. Kuala Lumpur. 429–444.

Raj, J.K. 1994. Characterization of the weathered profile developed over an amphibole schist bedrock in Peninsular Malaysia. *Bulletin of the Geological Society Malaysia*. 35: 135–144.

Singh, H. & Huat, B.B.K. 2004. Origin, formation and occurrence of tropical residual soils. Chapter 1 in *tropical residual soil engineering*. (eds) Huat et al. Balkema. UK: Taylor and Francis.

Stauffer, P.H. 1973. *Cenozoic in geology of the Malay Peninsular*. DJ Gobbett and CS Hutchinson (eds). Wiley Interscience.

Terzaghi, K. & Peck, R.B. 1967. *Soil mechanics in engineering practice*. 2nd ed., New York: John Wiley.

The Unesco Press, 1976. Engineering geological maps: A guide to their preparation. Paris: Unesco Press.

Varnes, D.J. 1978. Slope movement types and processes, *in* RL Schuster and RJ Krizek (ed.), *Landslides – Analysis and control*: National Academy of Sciences, Transportation Research Board Special Report (176): 12–33.

West, T.R. 1995. *Geology applied to engineering*. Prentice Hall.

Yogeswaran, M. 1995. Geological considerations in the development of the Kuching area. *Proceedings of the Dialogue Session on Geological and Geotechnical Considerations in Civil Works*. Geological Survey of Malaysia. Kuala Lumpur. 59–101.

CHAPTER 3

Site Investigation, Characterization and Selection of Design Parameters for Foundation Design

Ting Wen Hui, R. Nithiaraj & Tan Ek Hai
Zaidun-Leeng Sdn. Bhd., Kuala Lumpur, Malaysia

3.1 INTRODUCTION

Ground strength, stiffness, permeability and ground water profile are essential design parameter inputs required in the assessment and decision process for the selection of a suitable foundation system within a project site. Other relevant inputs are existing topography, proposed land use, proposed platform levels, loading details and information on foundation of adjacent development to study effects and consideration of temporary and permanent works. In this chapter, the planning of site investigation and processes involved in the characterization of collected data that leads to the selection of design parameters shall be described.

3.2 SITE INVESTIGATION

Objectives

The main objectives of site investigation are:

i. Assess the general suitability of the site and environments for the proposed works;
ii. Enable preparation of safe and economic design for both and temporary and permanent works;
iii. Plan method of construction to overcome expected difficulties due to ground conditions;
iv. Explore borrow sources for material and disposal sites; and
v. Evaluate effects of works on adjacent property.

Stages of investigation

Site investigation generally proceeds in the stages as described below:

i. Desk study
   - land use maps
   - geological memoirs/maps
   - topography, admiralty charts, tide tables
   - local rainfall records
   - past site investigation records of site and vicinity
   - past construction records
   - details and locations of existing services/utilities
   - aerial photographs
ii. Site reconnaissance
   - visual inspection of ground terrain/land use and check against that reported from desk study
   - condition of existing platform, slopes and structures and location/extent/type of distress/stabilization works if any
   - material type and signs of ground water seepage/wetness on exposed slopes

iii. Topographic/hydrographic investigation
   – survey of site and boundaries
   – details on type/location/dimensions/depths of existing structures/services
   – size and nature of catchment areas, location and details of streams and drainage
iv. Geological/hazard mapping
   – assess mass characteristics, distribution and structure of soil and rock (i.e. deposit type, origin, weathering profiles etc)
v. Sub-surface investigation:
   – identify problem areas and plan investigation type and extent to evaluate problems
vi. Dilapidation or pre-condition survey of surrounding properties/services prior to commencement of works.

Planning site investigation

Aspects of land development where geotechnical considerations are required are given below:

i. Evaluate adequacy in stability of embankment slopes, edges of fills; e.g. roads/expressways, platforms for infrastructure development etc; and provide recommendations.
ii. Evaluate adequacy of foundation support (i.e. bearing capacity and settlement) of structures, formed platform for structure and infrastructure facilities i.e. sewer pipes, water pipes, drains, roads etc., and provide recommendations.

To evaluate and assess the problems identified above, it is essential to carry out a site investigation to ascertain the character and variability of the strata underlying as well as surrounding the site of the proposed structure. Amongst the relevant considerations in the selection of investigation type and extent are:

– Knowledge of proposed land use, proposed finished levels, location and geometry of structures/loading. Identify problem areas for evaluation and plan investigation to gather relevant input data for evaluation.
– Select investigation type that is able to explore to sufficient depths and sample all strata likely to be affected by the structural load and construction operations. General guideline: exploration should be carried out to a depth where the net increase in soil stress under the weight of the structure is less than 10% of the average load of the structure or less than 5% of the effective stress in the soils at that depth, whichever is the lesser depth or unless bedrock is encountered first. Investigation of the bedrock is necessary where loads are to be carried to rock. Investigation should terminate when coring is continuous to at least 6 m competent rock below founding level to preclude possible boulder. In the case of deep excavations, investigations should be carried out to at least 1.5 times the depth of excavation, to locate and obtain piezometric levels in any aquifers that may exist below the level of excavation. This is to assist planning and evaluation of effects from dewatering.
– Determine the lateral extent and interval of investigation are sufficient to obtain information on type, consistency, thickness, sequence, discontinuities and dip of strata, and ground water conditions. Investigations are in general carried out in stages. At the initial stages, the investigation spacing is wide. At the detailed design stage when location and type of structure have been confirmed, the spacing of additional investigation is closer based upon the variability of the sub-surface condition from the initial stage investigations. For structures, 10 to 30 m spacing between investigation points is appropriate whilst for structures small in plan area, a minimum of 3 investigation points is recommended. Closer spacing is recommended at sites where engineering works are sensitive with difficult geological formations e.g. limestone.
– Provide for retrieval of samples (undisturbed and disturbed) to carry out laboratory tests to obtain index, stiffness, compressibility and strength parameters.

- Provide for in situ/field type tests that may more readily provide information for bearing capacity and settlement.
- An estimated cost and time required for the investigation works is equally important at the planning stage.

Subsurface investigation methods

Available methods for investigation may be intrusive (i.e. excavation, boring, probing) and non-intrusive (i.e. geophysical):

i. Excavation
Trial pits up to 5 m depths are dug manually or using an excavator. Retrieval of soil samples and in situ tests can be carried out at particular depths as pit is excavated. The exposed side slopes lend themselves to be mapped for geological structures.

ii. Borings
Shallow borings (<5 m) are carried out vide hand auger, sometimes with the use of light hand operated equipment. The boreholes are suitable in self-supporting strata without hard obstructions or gravel sized particles. Retrieval of soil samples and ground water observations can be carried out in the boreholes.

Deep borings (>5 m) are carried out vide rotary drilling methods. The method involves the rotation of a cutting shoe/drill bit located at tip of the drill rod to create the hole. Drilling fluid and/or casing are used to stabilize the formed part of the hole from collapsing in poor ground. Soil samples and field testing are usually carried out at 1.5 m depth intervals or a change in soil strata whichever is lesser. A typical drilling rig set up is shown in Figure 3.1.

iii. Probings
Probing methods commonly used are dynamic (e.g. Mackintosh Probes, which is also commonly known as JKR dynamic cone penetrometer, shown in Figure 3.2) or static (e.g. Cone Penetration Tests). Probes comprise sharpen steel rods that are driven (by hammer blow) or pushed into the soil (by static load). Investigation depth is limited by thrust available for pushing the penetrometer into the ground or when dense/coarse gravel is encountered. Soil type, strength and compressibility can be assessed from available correlation charts developed for the respective methods.

Figure 3.1.   Typical drilling rig set up.

Figure 3.2.   Mackintosh probe test.

Figure 3.3.   Preparation of CPT set up.

Probings can be completed relatively quicker compared to borings and are also more economical. In general, probings are used to complement boring results (i.e. interpolating ground profile between boreholes). Some preparatory measures prior to commencement of Cone Penetration Test are shown in Figure 3.3.

iv. Geophysical methods
Geophysical methods are used to obtain a preliminary assessment of the ground conditions. The results indicate variations and anomalies that can be correlated with geological or man-made features. This information is then used to interpolate ground conditions between boreholes or to indicate locations where further investigation by direct methods are required. Some techniques, applications and remarks are as in Tables 3.1 and Table 3.2 (BS5930 1999). Some preparatory measures prior to commencement of Geophysical Test are shown in Figure 3.4.

Table 3.1.   Geophysical methods in ground investigation (BS5930 1999).

Problem		Example	Methods and remarks
Geological	Stratigraphical	Sediments over bedrock: i) Sands and gravel over bedrock, water table low in sands and gravels	*Land* Seismic refraction
		ii) Sands and gravels overlying clay, water table high in sands and gravels	Resistivity
		iii) Clay over bedrock	Resistivity or seismic refraction
		Sediments over bedrock normally	*Marine.* Continuous seismic reflection profiling
	Erosional (for caverns, see "Shafts…" below)	Buried channel	Seismic refraction Resistivity for feature wider than depth of cover
		Buried karstic surface	Resistivity contouring
	Structural	Buried faults, dykes	Resistivity contouring Seismic reflection and refraction Magnetic and gravimetric (large faults)
Resources buried	Water	Location of aquifer Location of saline/potable interface	Resistivity and seismic refraction
	Sand and gravel	Sand, gravel over clay	*Land.* Resistivity
		Gravel banks	*Marine.* Continuous seismic profiling, side scan sonar, echo sounding
	Rock	Intrusive in sedimentary rocks	Magnetic
	Clay	Clay pockets	Resistivity (weathering may give low resistivity)
Engineering parameters	Modulus of elasticity, density and porosity	Dynamic deformation modulus	Seismic velocity at surface, or with single or multiple boreholes (cross hole transmission) Borehole geophysics
	Depth of piles	Check on effects of ground treatment	
	Rock rippability	Choice of excavation method	Seismic
	Corrosivity of soils	Pipeline surveys	Surface resistivity. Redox potential
Buried artefacts	Cables	Trenches on land	Magnetometer Electromagnetic field detectors
	Pipes	Submarine trenches	Echo sounding, side scan sonar
		Submarine pipelines	Side scan sonar, magnetic, continuous seismic profiling (especially if thought to be partially buried) with high frequency pinger
	Shafts, adits and caverns	Shaft, sink holes, mine workings	Resistivity. Magnetics, electromagnetic, radar, infra-red air photography on clear areas Cross hole seismic measurements Detailed gravity for large systems
	Archaeological remains	Foundations, buried wall, crypts	Magnetic, electromagnetic resistivity and radar

Table 3.2.   Usefulness of engineering geophysical methods (BS5930 1999).

Geophysical method	Geotechnical applications									
	Depth to bedrock	Stratigraphy	Lithology	Fractured zones	Fault displacements	Dynamic elastic moduli	Density	Rippability	Cavity detection	Buried artefacts
Seismic										
– Refraction	4	4	3	3	4	3	2	4	1	1
– Reflection – land	2	2	2	1	2	0	0	0	2	1
– Reflection – marine	4	4	2	2	4	0	0	1	0	2
– Cross-hole	2	2	3	3	1	4	2	2	3	2
Electrical										
– Resistivity sounding	4	3	3	2	2	0	0	1	2	1
– Induced polarization (IP)	2	2	3	1	0	0	0	0	0	0
– Electromagnetic (EM) and resistivity profiling	3	2	2	4	1	0	0	0	3	3
Other										
– Ground probing radar	2	3	1	2	3	0	0	0	3	4
– Gravity	1	0	0	0	2	0	2	0	2	1
– Magnetic	0	0	0	0	2	0	0	0	2	3
Borehole										
– Self-potential	2	4	4	1	1	0	0	0	1	1
– Single point resistance	2	4	4	0	0	0	0	0	0	0
– Long and short normal, and lateral resistivity	2	4	4	0	0	0	0	0	0	0
– Natural gamma	2	4	4	0	0	0	0	0	0	0
– Gamma-gamma	3A	4	4	0	0	0	3A	0	0	0
– Neutron	2A	4	4	0	0	0	3A	0	0	0
– Fluid conductivity	0	1	0	0	0	0	1	0	0	0
– Fluid temperature	0	0	0	1	0	0	0	0	1	0
– Sonce (velocity)	3	4	2	3	0	3	2	1	2	0

KEY

0 = Not considered applicable; 1 = limited use; 2 = used, or could be used, but not best approach, or has limitations; 3 = excellent potential but not fully developed;
4 = generally considered an excellent approach, techniques well developed;
A = in conjunction with other electrical or nuclear logs.

Table 3.2. Usefulness of engineering geophysical methods (continued) (BS5930 1999).

Geophysical method	Geotechnical applications									
	Ground water exploration	Water quality	Porosity	Permeability	Temperature	Flow rate and/or direction	Buried channel	Clay pockets in limestone	Sand and gravel	Basic igneous dykes
Seismic										
– Refraction	2	0	0	0	0	0	4	1	2	1
– Reflection – land	2	0	0	0	0	0	1	0	0	1
– Reflection – marine	0	0	0	0	0	0	4	0	0	0
– Cross-hole	0	0	0	0	0	0	2	0	1	2
Electrical										
– Resistivity sounding	4	4	3	1	0	0	3	0	3	0
– Induced polarization (IP)	3	1	3	2	0	0	2	1	1	1
– Electromagnetic (EM) and resistivity profiling	4	4	1	0	0	0	3	4	3	3
Other										
– Ground probing radar	2	2	1	0	0	0	2	2	1	2
– Gravity	1	0	0	0	0	0	2	1	1	2
– Magnetic	0	0	0	0	0	0	1	3	0	4
Borehole										
– Self-potential	4	2	0	0	0	0	0	0	0	0
– Single point resistance	4	2	1	0	0	0	0	0	0	0
– Long and short normal, and lateral resistivity	4	2	4	0	0	0	0	0	0	0
– Natural gamma	2A	2	1A	3A	0	0	0	0	0	0
– Gamma-gamma	2A	0	3A	2A	0	0	0	0	0	0
– Neutron	2A	0	3A	2	0	0	0	0	0	0
– Fluid conductivity	4	4	4	1	0	0	0	0	0	0
– Fluid temperature	2	3	0	0	4	2	0	0	0	0
– Sonce (velocity)	1	0	1	0	0	0	0	0	0	0

KEY
0 = Not considered applicable; 1 = limited use; 2 = used, or could be used, but not best approach, or has limitations; 3 = excellent potential but not fully developed;
4 = generally considered an excellent approach, techniques well developed;
A = in conjunction with other electrical or nuclear logs.

Figure 3.4.   Geophysical test with digital readout unit.

Figure 3.5.   Split tube sampler for obtaining samples from SPT tests.

Sample and the techniques

The four main techniques to retrieve samples are:

  i. Using drill tools or excavation equipment in the course of boring or excavation;
 ii. Using a tube or split tube sampler forced into the ground either by a static thrust or by dynamic impact (Figure 3.5);
iii. Using a tube with a cutter rotated into the ground; and
 iv. From block samples cut by hand from the trial pit.

Samples retrieved are often grouped to reflect sample disturbance. Sample disturbance is related to the sampling procedure or technique adopted. Tabulation on expected sample quality from different sampling procedures and soil sample quality classification are as found in Tables 3.3 and Table 3.4 (GEO 1993).

Table 3.3. Expected sample quality from different sampling procedures for (GEO 1993).

Material type	Typical composition of material	Sampling procedure	Expected quality class
Soils derived from in situ rock weathering	Composition of soils varies depending on the nature of parent rock material. Soils derived from granitic rock are usually silty and clayey sands; soils derived from volcanic rock are usually sandy and clayey silts	Block sample from dry excavation	1
		Large diameter triple-tube core-barrel (102 mm diameter cores) with retractor shoe, air-foam flush	1
		Triple tube core-barrel (≥74 mm diameter cores) with retractor shoe	1/2
		U100 sampler	2/4
		SPT split barrel sampler with or without liner	3/4
		Bulk samples and jar samples from dry open excavation	3/4
		Light percussion shell and chisel for boulders	5
Colluvium	Fresh or variably decomposed rock fragments (boulders, cobbles and gravels) within a matrix of varying proportions of sand, silt and clay	The sampling procedures for soils derived from in situ rock weathering apply	
Alluvial and marine deposits	The following materials can be found:		
	(a) Granular soils (sands, silty sands or sandy silts)	Piston sampler or compressed air sand sampler	2/3
		U100 sampler (with core-catcher)	4
		SPT split barrel sampler	4
		Light percussion shell	5
	(b) Very soft to soft cohesive soils (sandy clays, silty clays or clays)	Piston sampler	1
		Thin-walled sampler	1/2
		U100 sampler	2/3
		Delft continuous sampler	2/3
		Light percussion clay cutter (dry boreholes) or shell (wet boreholes)	4/5
	(c) Firm to very stiff cohesive soils	Triple-tube core-barrel with retractor shoe	1/2
		U100 sampler	2/3
		Light percussion clay cutter	5
	(d) Cohesive and granular soils containing boulders, cobbles or gravel	The sampling procedures for soils derived from in situ rock weathering apply	
Fill	Variable material, which can include compacted or uncompacted soil, rock fragments and building debris mixtures	See sampling procedures for relevant soil type and composition under "Alluvial and Marine Deposits" above.	
Rock	All rock types found in Hong Kong, including boulders in colluvium	Diamond core drilling with double or triple-tube core-barrel. The latter generally causes less disturbance and gives better core recovery, especially in highly fractured or jointed rocks.	N/A

Notes: The typical composition of materials should only be taken as a general guide. (2) The quality classes given should only be taken as a guide, as sample quality is highly dependent on workmanship and on the compactness (or consistency) and grading of the soil.

Table 3.4.   Soil sample quality classification (from GEO 1993).

Sample quality	Soil properties that can be reliably determined
Class 1	Classification, moisture content, density, strength, deformation and consolidation characteristics
Class 2	Classification, moisture content, density
Class 3	Classification, moisture content
Class 4	Classification
Class 5	None (approximate sequence of materials only)

Notes: (1) Large diameter Class 1 and Class 2 samples are often sufficient to allow the "fabric" of the soil to be examined. Sometimes this may also be done using Class 3 and Class 4 samples. (2) Remoulded properties can be obtained using Class 1 to Class 4 samples. (3) Table taken from BS 5930 (BSI, 1981a).

Field tests

Field test are carried out in the excavated pits or boreholes at the required depths. The results reflect the mass characteristics of the ground as well as overcome disturbance effects of samples retrieved for laboratory tests. Common field tests carried out to assess strength, compressibility and seepage parameters are as presented in Table 3.5 (Canadian Geotechnical Society 1987). Typical equipment for pressuremeter test and test setup for plate bearing test are shown in Figures 3.6 and 3.7 respectively.

Laboratory tests

Laboratory tests are carried out to

  i. classify samples,
 ii. investigate the fundamental behavior of the soils and rocks to determine the most appropriate method for analysis, and
iii. obtain soil and rock parameters relevant to the technical objectives of the investigation and to complement field test results.

The required mass of sample for the various laboratory tests are as presented in Table 3.6 (BS5930 1999). Common laboratory tests on soils and ground water with remarks on their use are presented in Table 3.7 (BS5930 1999). Corresponding laboratory tests on rocks are as shown in Table 3.8 (BS5930 1999). Figures 3.8 and 3.9 show typical laboratory test set-ups for undisturbed soil samples.

Recording and reporting

Data and test results collected in the field and laboratory are compiled into a report. Relevant data to be included in the report are given below:

• Date of investigation and an identification number to reference investigated location e.g. BH 1 etc
• Existing ground level at investigated location that is referenced to an established datum
• Location of investigated point (i.e. coordinates) referenced to an established state or national grid
• Drawing showing location of investigated locations with sufficient topographical details
• Date of testing, sample location and depth retrieved
• Methods of investigation and testing, equipment used, date commenced and date completed
• Investigation logs that describe material type (both engineering and geological) with depth
• Ground water levels where encountered should be reported
• Field test results in summary and detail
• Laboratory test results in summary and detail.

Table 3.5. Common in-situ tests (from Canadian Geotechnical Society 1987).

Type of test	Best suited to	Not applicable to	Properties that can be determined	Remarks	References
Standard penetration test (SPT)	Sand	Soft to firm clays	Qualitative evaluation of compactness Qualitative comparison of subsoil stratification		ASTM D 1586-67 Peck et al. (1974) Tavenas (1971) Kovacs et al. (1981) ESOPT II (1982) Schmertmann (1979) BS 5930 (1999) – C1 25.2
Dynamic cone test	Sand and gravel	Clay	Qualitative evaluation of compactness Qualitative comparison of subsoil stratification		ISSMFE (1977b) Ireland et al. (1970) BS 5930 (1999) – C1 26.2 Ooi & Ting (1975)
Static cone test	Sand, silt and clay		Continuous evaluation of density and strength of sands Continuous evaluation of undrained shear strength in clays	Test is best suited for the design of footings and piles in sand; tests in clay are more reliable when used in conjunction with vane tests	Sanglerat (1972) Schmertmann (1970, 1978) ESOPT II (1982) ASTM D 3441-79 Robertson and Campanella (1983a,b) Schapp and Zuidberg (1978) BS 5930 (1999) – C1 26.3
Field vane test	Clay		Undrained shear strength	Test should be used with care, particularly in fissured, varved and highly plastic clays	ASTM D 2553-72 Bjerrum (1972) Schmertmann (1975) Wroth and Hughes (1973) Wroth (1975) BS 5930 (1999) – C1 25.3

(Continued)

Table 3.5. Continued.

Type of test	Best suited to	Not applicable to	Properties that can be determined	Remarks	References
Pressuremeter test	Soft rock, sand, gravel and till	Soft sensitive clays	Bearing capacity and compressibility		Menard (1965) Eisenstein and Morrisson (1973) Tavenas (1971)
Plate bearing test and Screw plate test	Sand and clay		Deformation modulus Modulus of subgrade reaction Bearing capacity	Strictly applicable only if the deposit is uniform; size effects must be considered in other cases	ASTM D 1994-72 BS 5930 (1999) – C1 25.6
Flat plate dilatometer test	Sand and clay	Gravel	Empirical correlation for soil type, K_0, overconsolidation ratio, undrained shear strength, and modulus		Marchetti (1980) Campanella and Robertson (1982)
Permeability test	Sand and gravel		Evaluation of coefficient of permeability	Variable-head tests in boreholes have limited accuracy. Results reliable to one order of magnitude are obtained only from long term, large scale pumping test	Hvorslev (1949) Sherard et al. (1963) Olson and Daniel (1981) Tavenas et al. (1983a, b) BS 5930 (1999) – C1 25.4

* nb: References in table are not included in reference list.

Figure 3.6.   Menard pressuremeter test equipment.

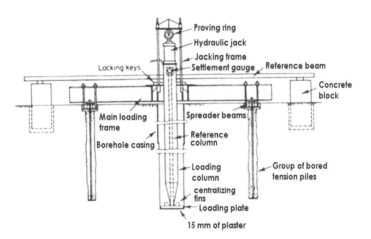

Figure 3.7.   Plate bearing test set up.

Table 3.6.   Mass of soil required for various laboratory tests (BS5930 1999).

Purpose of sample	Soil type	Mass of sample required (kg)
Soil identification, including Atterberg limits; sieve analysis; moisture content and sulphate content tests	Clay, silt, sand Fine and medium gravel Coarse gravel	1 5 30
Compaction tests	All	25 to 60
Comprehensive examination of construction materials, including soil stabilization	Clay, silt, sand Fine and medium gravel Coarse gravel	100 130 160

Table 3.7. Common laboratory tests for soil (from BS5930 1999).

Category of test	Name of test or parameter measured	Where details can be found	Remarks
Classification tests	Moisture content or water content	BS1377 1990 – Part 2	Frequently carried out as a part of other soil tests. Read in conjunction with liquid and plastic limits, it gives an indication of undrained strength
	Soil suction	Chandler et al. (1992)	To assess negative pore pressure in samples; especially for desiccated soils
	Liquid and plastic limits (Atterberg Limits)	BS1377 1990 – Part 2	To classify fine grained soil and the fine fraction of mixed soil
	Volumetric shrinkage limit	BS1377 1990 – Part 2	To determined the moisture content below which clay ceases to shrink
	Linear shrinkage	BS1377 1990	To assess the magnitude of shrinkage on desiccation
	Swelling clay content	BS1377 1990	Relevant to expansive materials and based on total cation exchanged capacity of soil
	Particle density	BS1377 1990 – Part 2	Values commonly range between 2.55 and 2.75 but a more accurate value is required for air voids determination. Only occasional checks are needed for most British soils, for which a value of 2.65 is assumed unless experience of similar soils shows otherwise
	Mass density or unit weight	BS1377 1990 – Part 2	Used in the calculation of forces exerted by soil
	Particle size distribution (grading)	BS1377 1990	Sieving methods give the grading of soil coarser than silt and the proportions passing the finest sieve represents the combined silt/clay fraction
	a) sieving	BS1377 1990 – Part 2	
	b) sedimentation	BS1377 1990 – Part 2	When the sample contains silt or clay, the test should be done by wet sieving. The relative proportions of silt and clay can only be determined by means of sedimentation tests
Chemical and electro-chemical tests	Dispersion	BS1377 1990 – Part 5	Qualitative tests to assess the erodibility of fine grained soils
	Contaminants	BS5930 1999 – annex F	This is a rapidly developing field; check the most recent guidelines
	Organic matter	BS1377 1990 – Part 3 BS 1924	Detects the presence of organic matter able to interfere with the hydration of Portland cement in soil: cement pastes
	Mass loss on ignition	BS1377 1990 – Part 3	Measures the organic content in soils, particularly peats
	Sulfate content of soil and ground water	BS1377 1990 – Part 3	Assesses the aggressiveness of soil or groundwater to buried concrete (See remarks on test for the pH value and chloride content)
Chemical and electro-chemical tests	Magnesium content	Bowley (1979)	Supplements the sulfate content test to assess the aggressiveness of soil or groundwater to buried concrete
	pH value	BS1377 1990 – Part 3	Measures the acidity or alkalinity of the soil or water. It is usually carried out in conjunction with sulfate content tests. This test and the two above should be performed as soon as possible after the samples have been taken
	Carbonate content	BS1377 1990 – Part 3	Confirms the presence of carbonates, which often indicates cementing

	Chloride content	BS1377 1990 – Part 3	Test recommended where pH of ground is less than 5.8. Results used in conjunction with those for sulfate, nitrate and pH to assess aggressiveness of ground, especially to concrete
	Total dissolved solids in groundwater	BS1377 1990 – Part 3	A general measures of salinity indicative of aggressiveness of ground and related to electrical conductivity or soil resistivity
Soil corrosivity tests	a) Bacteriological	BS 7361-1	Undisturbed specimens required in sterilized containers
	b) Redox pot	BS 7361-1	
		BS1377 1990 – Part 3	
	c) Resistivity	BS1377 1990 – Part 3	
Compaction-related tests	Dry density (or dry unit weight)	BS1377 1990 – Part 9	Measure the mass (or weight) of solids per unit volume of soil. Often used as a quality control for compaction of fill
	Standard compaction tests	BS1377 1990 – Part 4	Indicate the degree of compaction that can be achieved at different moisture contents with different compactive efforts
	Maximum, minimum density and density index of coarse grained soil	BS1377 1990 – Part 4	Density index indicates the stiffness and peak strength of coarse grained soils. A number of different methods are available, so the method used should be clearly stated
	Moisture condition value (MCV)	BS1377 1990 – Part 4	Determines compactive effort required to produce near-full compaction. Used for control of materials for earthworks
Pavement design tests	California Bearing ratio (CBR)	BS1377 1990 – Part 4	This is an empirical test used for design of flexible pavements. The test can be made either in situ or in the laboratory
	Chalk crushing value	BS1377 1990 – Part 4	Similar in concept to the aggregate crushing value (ACV)
	Frost heave test	BS812	Assesses susceptibility of compacted soil to frost heave
	Aggregate suitability	BS812	Physical and chemical tests for aiding the selection and assessing the suitability of materials to act as bound and unbound aggregates
Soil strength tests	Triaxial compression:	BS1377 1990 Head (1986)	Triaxial tests are normally carried out on nominal 100 mm or 38 mm diameter samples with height to diameter ratio 2:1. If the height to diameter ratio is reduced to 1:1 the end platens should be lubricated Undrained tests measure undrained strength s_u. Drained tests, or undrained tests with measurement of pore pressure, evaluate the Mohr Coulomb parameters c' and ϕ'. Since soil strength depends on strain it is necessary to state whether the strength corresponds to the peak state, the critical state or the residual (Atkinson 1993)

(Continued)

Table 3.7. Continued.

Category of test	Name of test or parameter measured	Where details can be found	Remarks
	a) Unconsolidated undrained	BS1377 1990 – Part 7	Prior to triaxial shearing, samples may be consolidated in the apparatus to some specified
	b) Undrained with measurement of pore water pressure	BS1377 1990 – Part 8	state: these are then known as consolidated undrained or consolidated drained tests as appropriate. Any drained or undrained test in which port pressures are measured should
	c) Undrained with measurements of volume change	BS1377 1990 – Part 8	be consolidated before shearing
	d) Multi-stage	BS1377 1990 – Part 7	Several techniques have been used for both drained and undrained multi-stage tests, details of which may be found in the references. The test may be useful where there is a shortage of specimens. Multi-stage tests are not recommended when single stage tests can be carried out
	e) Stress path test	Bowley (1979)	Stress paths other than those used in a) to c) may be applied to reproduce the history of stress and strain in the ground before and during construction
	Unconfined compression test	BS1377 1990 – Part 7	This simple test is a rapid substitute for the undrained triaxial test. It is suitable only for saturated non-fissured fine grained soil
	Laboratory vane shear	BS1377 1990 – Part 7	For soft clay, as an alternative to the undrained triaxial test or the unconfined compression test
	Direct shear box	BS1377 1990 – Part 7	Direct shear tests are an alternative to triaxial tests although the latter are more versatile and more often used. Disadvantages are: drained conditions cannot be controlled nor pore pressures measured and the plane of shear is predetermined by the nature of the test. An advantage is that samples of coarse grained soil can be more easily prepared than in the triaxial test. In general only drained tests should be undertaken. Shear boxes are normally square with sides 60 mm or 100 mm but may also be circular in plan. For very coarse grained soils shear boxes with sides 300 mm or larger should be used
	Residual shear strength: a) Multiple reversal shear box b) Triaxial test with pre-formed shear surface c) Shear-box test with preformed shear surface d) Ring shear test	BS1377 1990 – Part 7	The residual shear strength of clay soil is relevant for slope stability problems where previous sliding has developed residual slip planes in situ. The multiple reversal shear box test is the one that is most commonly used, although the ring shear test would be the more logical choice
Soil deformation tests	One-dimensional and compression and consolidation tests:	BS 1377	These tests measure soil parameters m_v and c_v for simple calculations of the magnitude and rate of settlement of foundations

	a) Standard (incremental loading) oedometer test	BS1377 1990 – Part 5	The standard dead weight loading oedometer is the one in general use
	b) Continuous loading oedometer tests	Head (1986) Atkinson (1990)	The alternative is the hydraulic oedometer (Rowe cell) in which the vertical loading and the pore pressures can be independently controlled Reasonable assessments of the magnitudes of foundation settlements can be made if
	c) Swelling and collapse on wetting	BS1377 1990 – Part 5	Class 1A samples are tested For stiff clay, a careful load-unload and reload sequence is applied using small increments and decrements For soft clay, reliable determinations of the yield shell are made Estimates of settlement can be much improved if small strain triaxial and pressuremeter tests are used Estimates the rate of settlement that have been found to be highly inaccurate with certain types of soil Instead of applying the loads in discrete increments, as in the standard test, stresses, strains or pore pressures may be varied continuously Additional tests are carried out to determine the swelling pressure and the swelling or settlement on saturation
	Shear and bulk modulus	Atkinson (1991)	Stress/strain relations for soils are highly non-linear and the bulk modulus and shear modulus both vary with loading. For the relatively small loadings, appropriate to most engineering applications, soil strains are relatively small (typically less than 0.1%) Measurement of these small strains requires use of special apparatus and procedures. These include use of local strain gauges attached to the sample and application of stress paths closely resembling the field stress paths
Soil permeability tests	Tests in permeameters a) Constant head test	BS 1377	The constant head test is suited only to soils of permeability normally within the range 10^{-4} m/s to 10^{-2} m/s. For soils of lower permeability, the falling head test is applicable
	b) Falling head test Triaxial permeability test Rowe consolidation cell	Head (1982) BS1377 1990 – Part 6 BS1377 1990 – Part 6	For various reasons, laboratory permeability tests often yield results of limited value and in situ tests are generally thought to yield more reliable data The triaxial cell and the Rowe consolidation cell allow the direct measurement of permeability under constant head with a back pressure and confining pressures more closely consistent with the field state. The Rowe cell allows either vertical or radial flow

* nb: References in table are not included in references list.

Table 3.7. Common laboratory tests for rock (from BS5930 1999).

Category of test	Name of test or parameter measured	Where details can be found	Remarks
Rock classification tests	Saturation moisture content (alteration index) Bulk density Moisture content Porosity	BS 812-2 ISRM (1979)	The parameters from these tests may be related to other parameters such as compressive strength, modulus of elasticity, seismic wave velocity, resistance to weathering and degree of weathering
	Petrographic analysis	ISRM (1978)	Useful to identify rock type and degree of weathering and gives an indication of stress history
	Slake-durability	ISRM (1979)	Useful quality index for testing clay-bearing rocks proposed for construction materials
	Hardness and abrasiveness	ISRM (1978)	Gives some indication of potential wear and tear of machinery involved in rock cutting and breaking
	Carbonate test	BS 1881-6	Reference describes a method using Collins' calcimeter. Useful for the identification of chalk and calcareous mudrocks
	Swelling test	BS 1377 ISRM (1979) ISRM (1989)	Gives some indication of moisture sensitivity of rock and possible ranges of induced pressures on tunnel linings
Dynamic tests	Seismic velocity Dynamic modulus	ISRM (1978) ISRM (1978) ISRM (1978)	Results may sometimes be useful in extrapolating laboratory and field tests to rock mass behavior
Rock strength tests	Point load test Uniaxial compression	ISRM (1985) ISRM (1979)	Simple laboratory and field strength test. Useful aid to core logging Carried out on intact samples with no discontinuities and yields data on the rock material properties. The length to diameter ratio of 2:1 is a minimum for cylinders
	Direct tension test Indirect tensile strength test (Brazil test)	ISRM (1978)	See remarks on uniaxial compression test

Test	References	Remarks
Triaxial compression: a) Undrained b) Undrained with measurement of pore water pressure c) Drained	ISRM (1983) ISRM (1989) HOEK et al. (1968) NEFF (1966) ISRM (1981)	Usually carried out on intact samples with no discontinuities and yield data on rock material properties. Larger samples may contain one or more discontinuities, in which case the data relates to the properties of the rock mass
Discontinuity strength tests — Direct shear box	KRSMANOVIC (1967) HOEK (1974) ISRM (1981)	Of considerable importance for study of friction on discontinuities. References also cover residual shear strength. Where joints are filled with gouge, the properties of the combined gouge and joint should be determined under conditions closely simulating those existing in situ
Rock deformation tests — Static elastic modulus	ISRM (1979) ISRM (1983)	See remarks on triaxial compression test
Creep tests: a) Undrained b) Constant load c) Triaxial	MEIGH et al. (1973) GRUDEN (1971)	Most meaningful when carried out under multi-stress conditions
Consolidation of rock mass containing gouge material		Treat the gouge material as a soil
Rock permeability tests — Triaxial cell test	Geological Society Engineering Group Working Party Report (1970)	Makes use of a modified Hoek-Franklin cell
Centrifugal test Radial test	Daw (1971) Hoek et al. (1974) Worthington (1972) Bernaix (1967)	Considerably faster than other methods A measure of the degree of fracturing of the rock material

*nb: References in table are not included in references list.

Figure 3.8.   Setting up a trixial test.

Figure 3.9.   Consolidation test set up.

3.3   CHARACTERIZATION PROCESS

Characterization process following data collection described in section herein before, involves the following stages:

 i. Data assembly
 ii. Data processing
iii. Data interpretation.

Data assembly

The collected data from subsurface investigation are assembled in formats that allow spreadsheet processing. Data to form the below plots may be extracted and assembled:

- SPT 'N' vs depth
- Particle size distribution (i.e. clay content percent vs depth etc)

- Moisture content, Atterberg Limits vs depth
- Vane shear vs depth
- Pressuremeter modulus vs depth
- m_v vs log pressure
- c_v vs log pressure
- Permeability vs depth
- Organic content vs depth
- pH vs depth
- Mackintosh blow count vs depth
- Etc.

Data processing

Inserting the assembled data in a spreadsheet, plots of the same are then created. From these plots or combinations thereof, relationships, if any, can then be observed, such as

- Trends of increase, decrease, no change with depth.
- Range of magnitudes with depth, SPT 'N' profile, weathering grade etc.
- Superimposition of plots on cross- sections or long sections to assess variation with location, depth and extent.

The above plots may be assembled considering geological formation e.g. alluvial, man made fill, residual soil, lithology type, weathering grades, topography etc.

3.4   SELECTION OF DESIGN PARAMETERS

Following the preceding section, relevant engineering properties of interest i.e. strength and stiffness/compressibility profile and in some cases the permeability profile may be selected at this stage.

Characterized strength values for design purposes are normally taken as the lower bound margin from plots created. Finally an engineering judgment has to be exercised to select the appropriate characteristic after having applied the following procedures:

i. The evaluation of the limitations of the relevant computation models
ii. The application of the principles of statistics in a practical manner
iii. The acquisition of the knowledge of the performance of similar constructions in similar formations and an awareness that method of construction has to be taken into consideration.

Characterization of in situ sediments, weathered formation, and engineered fills are further described below.

In situ sediments

Undrained strength ratio has been widely used as an empirical tool for soft soil. Wroth (1984) presented the interpretation of in situ soils within the general framework of the critical state soil mechanics and the deduced Cam clay model. He presented for engineering application the fundamental basis of the relationship:

$$S_u = S_{ru}P'_o \tag{3.1}$$

where S_u is the undrained shear strength equal to half the deviator strength, P'_o is the effective vertical stress on the tested sample, S_{ru} is the undrained strength ratio that has a particular value depending on the over-consolidation ratio and nature of the deposit of the sample. The simple relationship proposed is a powerful analytical tool for characterization. The S_{ru} values usually range between 0.2 and 0.25 for normally consolidated clayey soils. Higher values of S_{ru} will denote that a soil is over-consolidated. Strength characterization has been described by Ting (2003).

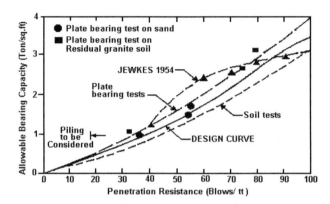

Figure 3.10.   Allowable bearing capacity versus JKR dynamic cone penetration resistance (Ooi & Ting 1975).

Weathered formation

Where a suitable computation model is not easily available, empirical procedures are often employed. A rational empirical method is nonetheless based on observed physical processes that realistically describe the problem. The empirical methods are closely related to load testing, measurement and monitoring of full-scale structures in what is known as observational methods. In geotechnical engineering, observational methods not only serve the design review process, but also provide the database to be applied to similar site problems in the future as well as for further development of theoretical methods.

In their natural state, residual soils exhibit different behavior partly due to grading, index properties, mineral compositions, structure and bonding inherited from the parent rock. In the case of excavated material for use as fill, the effects of structure and bonding may be less pronounced. The preceding sections outline process and tests methods used in the characterization of residual tropical soils as natural foundations and engineered fills.

Of engineering interest as a natural foundation are properties related to assess safe bearing capacity and settlement. In view of the inherent variation in behavior with lithology and weathering profiles for reasons described in the sections herein before, weathering profiles within the project site are required to be established prior to association with the respective geotechnical parameters. Deep boreholes distributed geographically and topographically with an Engineering Geologist/Geologist to log retrieved samples are recommended. Preparation of a project specific Surface Geology Map can also be of assistance. Methods and tests to obtain geotechnical parameters are briefly outlined below.

For slope bodies, the empirical approach of Bulman (1967) is relevant. With this scheme, typical slope angles may be recommended based on observations of similar slopes. A deviation from the general scheme would be investigated as a localized problem to be dealt with on a case-to-case basis for the given site during the slope formation construction process.

For shallow foundations such as footings, design curve on allowable bearing capacity from studies carried out with Mackintosh (JKR) probe by Ooi & Ting (1975) is available and reproduced in Figure 3.10.

For pile foundations such as bored piles, design parameters studies such as conducted by Tan et al. (1998) is available. Design curve relating shaft resistance factor with SPT 'N' value is reproduced in Figure 3.11. Pile base resistance from their studies is negligible and is recommended not be considered.

Laboratory test

The engineering properties required for the assessment of bearing capacities based upon engineering expressions developed by respected experts i.e. Vesic, Meyerhof, Terzaghi etc. are related to undrained and drained strength parameters i.e. c, ϕ, c', ϕ'.

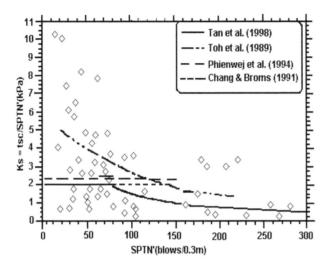

Figure 3.11.   Variation of K_s with SPT 'N' (Tan et al. 1998).

These strength parameters may be obtained vide strength envelopes determined from triaxial tests carried out on undisturbed samples. The portion of the envelope used for design purposes must reflect that for the desired design stress range. Samples selected for testing should be sufficient to cover the full weathering profile for the different lithologies.

Field test

Direct measurements on settlement and bearing capacity may be carried out vide plate bearing tests or pressuremeter tests conducted in boreholes at the required depths.

Engineered fill

Fill material may be either engineered, i.e. manner of deposition and compaction is controlled/ checked or non-engineered i.e. loosely dumped. Engineered fill embankments that are very high (i.e. >20 m) are prone to collapse type settlements upon saturation and require evaluation (Ting 1979, Ting & Chan 1991). Non-engineered fills do not lend themselves to be characterized as the material type and manner of deposition can be variable in extent as well as depth. Methods and tests to obtain geotechnical parameters are briefly outlined below.

The standard of improvement required for engineered fill is commensurate with the desired performance (Ting, 1999) of the end product: in providing a flood free surface, support of infrastructure or in more demanding circumstances in the support of superstructures (Ting & Chan 1991). As an example, for applications such as in drainage and irrigation bunds, conditioning of the in situ wet fill by drying out may be adequate for the performance required of low embankment and of imperviousness.

The quality of source of fill in the local context is a function of the content of granular material in terms of size and quantum (Ting & Ooi 1976, Ting et al. 1982). For clayey material at source, the natural moisture content is relevant.

The required compaction characteristics (maximum dry density and optimum moisture contents) are usually a function of the granular content and the moisture content. Disturbed samples obtained from source may be re-constituted, based on results of compaction tests, and be tested for shear strength.

For an appropriate sampling from source, an understanding of the geology of the deposits is essential.

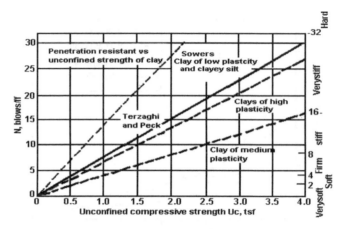

Figure 3.12.   Correlations of SPT N values with U_c for cohesive soils of varying plasticities. Source: NAVFAC Manual DM 7 (1971) as reported in Hunt (1984).

Laboratory test

Strength and compressibility parameters may be assessed from triaxial and consolidation tests results carried out on extruded samples that have been reconstituted to specified compaction requirements i.e. proctor test requirements. Collapsed type settlement may be estimated by repeating the consolidation test on saturated samples for each load increment.

Field test

Similar to the above section, plate bearing tests or pressuremeter tests carried out in boreholes are applicable for the assessment of bearing capacity and settlement. For initial reference to assess engineering properties of engineered fill, empirical chart and table in NAVFAC (1971) manual publication reproduced in Figure 3.12 and Table 3.9 may be of assistance.

3.5   CASE HISTORY EXAMPLES

Case history no.1 – weathered formation

Nithiaraj et al. (1996) reported that correlations between strength parameters and SPT 'N' values carried out on residual soils of granite and metasedimentary origin at Kampar, Perak, Malaysia, are shown to exhibit good linear relationships (Figures 3.13, 3.14, 3.15 & 3.16).

   The results were used to select strengths of weathered soils in the stability analysis.

Case history no.2 – in situ sediments

Ting et al. (2005) reported on a case of investigation carried out in Quaternary deposits at Slim River Perak, Malaysia. Figure 3.17 below shows the characterized strength profile of the ground. The characterized strength profiles shown below were based on Cone Penetrometer Test (CPT) and Field Vane Shear (FVS) test results and limited to depths of 15 m.

   Table 3.10 below shows the typical soil classification results for the three different geologies of organic soil, coastal and river alluvium.

   Pressuremeter tests (PMT) were also carried out and the results obtained were shown in Figure 3.18. Based on the PL results above, the undrained shear strength, S_u of the ground was determined following Amar et al. (1991). Together with the S_u determined from Field Vane Shear (FVS) and Cone Penetration Test (CPT) for the same site, some statistical analysis was carried out to compare the scatter of the results as shown in Table 3.11.

Table 3.8. Typical properties of compacted soils.

				Typical properties of compacted soils*								
				Typical value of compression		Typical strength characteristics						
				Percent of original height								
Group symbol	Soil type	Range of maximum dry unit weight, pcf	Range of optimum moisture, %	At 1.4 tsf (20 psi)	At 3.6 tsf (50 psi)	Cohesion (as compacted) psf	Cohesion (saturated) psf	Effective stress envelope ϕ, degrees	$\tan \phi$	Typical coefficient of permeability, ft/min	Range of CBR values	Range of subgrade modulus, k_s lb/in^3
GW	Well-graded clean gravels, gravel-sand mixtures	125–135	11–8	0.3	0.6	0	0	>38	>0.79	5×10^{-2}	40–80	300–500
GP	Poorly graded clean gravels, gravel-sand mix	115–125	14–11	0.4	0.9	0	0	>37	>0.74	10^{-1}	30–60	250–400
GM	Silty gravels, poorly graded gravel-sand silt	120–135	12–8	0.5	1.1	–	–	>34	>0.67	$>10^{-6}$	20–60	100–400
GC	Clayey gravels, poorly graded gravel-sand-clay	115–130	14–9	0.7	1.6	–	–	>31	>0.60	$>10^{-7}$	20–40	100–300
SW	Well graded clean sands, gravelly sands	110–130	16–9	0.6	1.2	0	0	38	0.79	$>10^{-3}$	20–40	200–300
SP	Poorly-graded clean sand, sand-gravel mix	100–120	21–12	0.8	1.4	0	0	37	0.74	$>10^{-3}$	10–40	200–300
SM	Silty sands, poorly graded sand-silt mix	110–125	16–11	0.8	1.6	1050	420	34	0.67	5×10^{-5}	10–40	100–300
SM-SC	Sand-silt clay mix with slightly plastic fines	110–130	15–11	0.8	1.4	1050	300	33	0.66	2×10^{-6}	–	

(Continued)

Table 3.9. Continued.

Group symbol	Soil type	Range of maximum dry unit weight, pcf	Range of optimum moisture, %	Typical value of compression / Percent of original height		Typical strength characteristics				Typical coefficient of permeability, ft/min	Range of CBR values	Range of subgrade modulus, k_s lb/in^3
				At 1.4 tsf (20 psi)	At 3.6 tsf (50 psi)	Cohesion (as compacted) psf	Cohesion (saturated) psf	Effective stress envelope ϕ, degrees	tan ϕ			
SC	Clayey sands, poorly graded sand-clay mix	105–125	19–11	1.1	22	1550	230	31	0.60	5×10^{-7}	5–20	100–300
ML	Inorganic silts and clayey silts, elastic silts	95–120	24–12	0.9	1.7	1400	190	32	0.62	10^{-5}	15 or less	100–200
ML-CL	Mixture of inorganic silt and clay	100–120	22–12	1.0	22	1350	460	32	0.62	5×10^{-7}	–	
CL	Inorganic clay of low to medium plasticity	95–120	24–12	1.3	25	1800	270	28	0.54	10^{-7}	15 or less	50–200
OL	Organic silts and silt-clays, low plasticity	80–100	33–21	–	–	–	–	–	–	–	5 or less	50–100
MH	Inorganic clayey silts, elastic silts	70–95	40–24	20	3.8	1500	420	25	0.47	5×10^{-7}	10 or less	50–100
CH	Inorganic clayey of high plasticity	75–105	36–19	26	3.9	2150	230	19	0.35	10^{-7}	15 or less	50–150

From NAVFAC Manual DM 7 (1971). All properties are for condition of 'Standard Proctor' maximum density, except values of k and CBR which are for 'modified Proctor' maximum density. Typical strength characteristics are for effective strength envelopes and obtained from USBR data. Compression values are for vertical loading with complete lateral confinement.
'–' indicates that insufficient data is available for an estimate.
(Source: NAVFAC Manual DM 7 (1971) as reported in Hunt 1984).

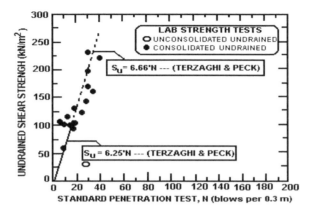

Figure 3.13.   Variation of undrained shear strength with SPT 'N' for weathered schist in grade V zone.

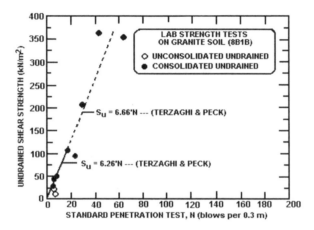

Figure 3.14.   Variation of undrained shear strength with SPT 'N' for weathered granite in grade V zone.

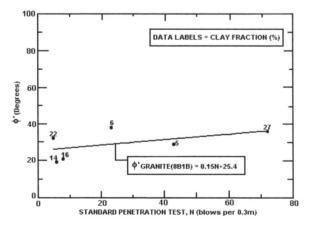

Figure 3.15.   Variation of ϕ' with SPT N for weathered granite.

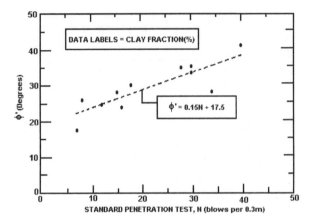

Figure 3.16.   Variation of ϕ' with SPT N for weathered schist.

| Soil type | Typical soil profile | | | Strength profile |
	Depth (m)	Description	S_U (kPa)	
Organic soil underlain by alluvium	0-2	Organic soil	8.0	
	2-4		3.0	
	4-6		10.0	
	6-8		13.0	
	8-10		15.0	
	10-12	CLAY	18.0	
	12-14		20.0	
Coastal alluvium	0-1	Marine CLAY (with seashells)	10.0	
	1-3		5.0	
	3-5		8.0	
	5-7		10.0	
	7-9		13.0	
	9-11		17.0	
	11-13		20.0	
	13-15		23.0	
River alluvium	0-3	Silty CLAY/ Clayey SILT	5.0	
	3-5		7.0	
	5-7		10.0	
	7-9		15.0	
	9-11		21.0	
	11-13		25.0	
	13-15		28.0	

Figure 3.17.   Characterized soil strength profile.

Table 3.10.   Typical soil classification for site.

| Soil type | NMC (%) | OC (%) | Particle size distribution (%) | | | | LL | PL | PI |
			Gravel	Sand	Silt	Clay			
Organic soil underlain by coastal alluvium	99	15	6	12	36	46	97	33	64
Coastal alluvium	90	11	0	15	34	51	96	40	56
River alluvium	57	12	6	38	26	30	81	33	48

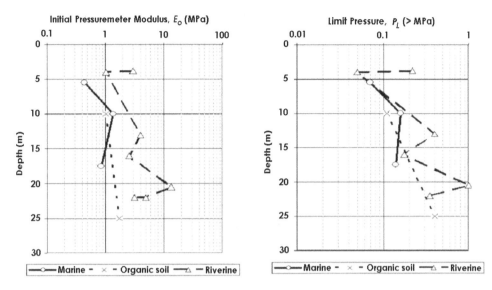

Figure 3.18.   Pressuremeter test (PMT) results.

Table 3.11.   Comparison of *Su* profile based on FVS, CPT and PMT.

	FVS	CPT	PMT
Mean, μ (kPa)	16.6	18.7	29.7
Coeff. of variance (COV)	0.61	1.10	1.39
Median (kPa)	14.5	8.4	24.4
Standarad deviation, σ(kPa)	10.1	20.6	41.4
Sample variance, σ^2	102	423	1,710
Minimum (kPa)	3.0	0.9	2.0
Maximum (kPa)	49.5	89.4	145.3
No. of tests	61	484	11

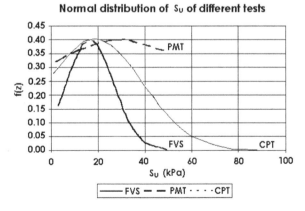

Figure 3.19.   Comparison of *Su* profile based on FVS, CPT and PMT.

The normal distribution curve presented in Figure 3.19 shows that the pressuremeter result has the greatest scatter and the mean value of the results collected was found to be nearly 2 times as much as the mean value for the CPT and FVS.

The usage of self-boring or displacement type pressuremeter may improve the results for tests carried out on soft ground. However, due to wider commercial availability, greater experience in usage (with good empirical correlation evident in the results shown above for the CPT) and economy of the other investigation techniques, the development of self-boring or driven type pressuremeter has not been forthcoming.

Yee (2004) has provided theoretical arguments as to why the FVS test is the preferred investigation method in investigating S_u in soft ground over other in situ methods such as the pressuremeter.

3.6 CONCLUSION

The processes and considerations involved in the planning of site investigation have been introduced. Considerations in the data assembly process that will lead to selection of design parameters have been described. Empirical charts and tables that are often relied upon in the practicing environment are provided.

REFERENCES

Amar S., Clarke B.G.F., Gambin M.P. & Orr T.L.L. 1991. The application of pressuremeter test results to foundation design in Europe. A state-of-the-art, report by *ISSMFE European Technical Committee on Pressuremeter*, Rotterdam: Balkema.

Bulman, J.N. 1967. A survey of road cuttings in western Malaysia. *Proceedings of the Southeast Asian Regional Conference on Soil Engineering*, Bangkok.

British Standard Institution, BS5930 1999: *Code of Practice for Site Investigations.*

Canadian Geotechnical Society 1987: *Canadian Foundation Engineering Manual.*

Geological Society Working Party 1990. Tropical residual soils, geological society engineering group working party report. *Quaterly Journal of Engineering Geology* 3(1).

GEO. 1993. Guide to site investigation: *Geoguide 2 Geotechnical Engineering Office*, Hong Kong.

Hunt, R.E. 1984. *Geotechnical Engineering Investigation Manual.*

NAVFAC Manual DM 7. 1971. Graphs and charts *Reproduced in Geotechnical Engineering Investigation Manual*, Roy E Hunt, 1984.

Nithiaraj, R., Ting, W.H. & Balasubramaniam, A.S. 1996. Strength parameters of residual soils and application to stability analysis of anchored slopes. *Geotechnical Engineering*, Dec 1996, SEAGS, Thailand. 27 (2).

Ooi, T.A. & Ting, W.H. 1975. The use of a light dynamic cone penetrometer in Malaysia. *Proc. 4th S.E. Asian Conference Soil Eng.* Kuala Lumpur.

Tan, Y.C., Chen, C.S., & Liew, S.S. 1998. Load transfer behaviour of cast-in place bored piles in tropical residual soils of Malaysia. *Proc. 13th S.E. Asian Geotechnical Conference*, Taipei.

Ting, W.H. & Ooi, T.A. 1972. Some properties of a Malaysian residual granite soil. *Third Southeast Asian Conference on Soil Engineering.*

Ting, W.H. & Ooi, T.A. 1976. Behaviour of a Malaysian residual granite soil as a sand-silt-clay composite soil. *Geotechnical Engineering, J of SEA Soc. of Soil Eng.*, December 1976. VII(2): 67–79.

Ting, W.H. 1979. Consolidation of a partially saturated residual soil. *Proc. 6th Asian Conf. on SM & Fn Engr.* Vol. 1, 1979, Singapore.

Ting, W.H., Mun, K.P. & Toh, C.T. 1982. Characteristics of a composite residual granite soil. *Proc. of 7th SEA Geot. Conf.*, Hong Kong.

Ting, W.H. & Chan, S.F. 1991. Improvement of fill by compaction for building purposes. *4th Asian Regional Conference on Soil Mechanics & Foundation Engineering*, Vol. 1, 9–13 Dec. 1991, Bangkok, Thailand.

Ting, W.H. 1999. A Survey of embankment construction practice and futured Developments. *Year 2000, Geotechnics*, AIT.

Ting, W.H. 2003. Charaterization of strength of in situ soil. Netherlands: A.A. Balkema, New Delhi: Oxford & IBH Publishing Co. Pte. Ltd.

Ting, W.H., Nithiaraj, R., Choong, P.K., Ang, K.A., & Tan, E.K. 2005. Some application of pressuremeter in Malaysia. *Proc. 5th International Symposium on Pressuremeter*, Paris.

Wroth, C.P. (1984). The interpretation of in situ soil tests. *Geotechnique*. 34(4): 449–489.

Yee T.S. 2004. Shear strength and structural stability of construction in soft ground. *Proc. Malaysian Geotech. Conf. 2004*, Kuala Lumpur: 475–495.

CHAPTER 4

Shallow Foundations

Faisal Hj. Ali
University Malaya, Kuala Lumpur, Malaysia

Bujang B.K. Huat
University Putra Malaysia, Malaysia

4.1 INTRODUCTION

For many of man's structures, it is the earth underlying the structure that provides the ultimate support. The soil at a building location automatically becomes a material of construction affecting the structure's stability. Typically, soil is a material weaker than other common materials of construction such as steel, concrete, and wood. To carry a given loading satisfactorily, a greater area or volume of soil is necessarily involved. For loads carried by steel, concrete, or wood structural members to be imparted to the soil, load transfer devices – the structural foundations – are required. The major purpose of the structural foundation is the proper transmission of building load to the earth in such a way that the supporting soil is not overstressed and does not undergo deformations that would cause serious building settlement. The type of structural foundation utilized is closely related to the properties of the supporting soils. A structural foundation performs satisfactorily only if the supporting soil behaves properly. Consequently, it is important to recognize that building support is actually being provided by a soil-foundation system, a combination that cannot be separated.

The type of foundation to be used for a structure or a building normally depends on the type of soil and the load imposed by the structure on the soil. A shallow foundation may be defined as one in which the embedment depth of the foundation is less than its least characteristic dimension. Foundations need to be capable of carrying an imposed loading without undergoing movement that causes structural damage or affects the facility's planned usage. These considerations require that the soil responsible for supporting a foundation is not stressed beyond its strength limits. Simultaneously, the deformations resulting within this soil because of loading and action of natural forces cannot be excessive. The magnitude and type of loading (static, live, or repetitive), the foundation performance requirements (how much settlement is permissible), and properties of the supporting soil, all influence the type and size of foundation that will be necessary and its resulting behavior.

4.2 TYPES OF SHALLOW FOUNDATIONS

Spread footings

Figure 4.1 shows some examples of commonly used shallow foundations. Spread or pad footing foundations are typically of plain concrete or reinforced concrete, although masonry and timber have also been used. The spread footing foundation (Figure 4.1 (a)) is basically a pad used to "spread out" the building columns and wall loads over a sufficiently large soil area. Spread footings are constructed as close to the ground surface as the building design permits (considering requirements such as basements or the need to resist lateral forces) and as controlled by local conditions (considering factors such as soil shrinkage and expansion, the possibility of soil erosion, or building code stipulations). Footings for permanent structures are rarely located directly on the ground surface. To be classified as a spread footing, the foundation does not have to be at a shallow depth; spread footings will be located deep in the ground if soil conditions or the building design requires it so. Spread footing foundations for building columns, walls, and equipment bases commonly have the shapes of squares, rectangles, trapezoids, and long strips (see Figure 4.1). Usually, the shape and dimensions for a footing result from having the structural loading positioned so that, theoretically, a *uniform* bearing pressure on the soil beneath the foundation is achieved. For the support of walls and single columns, the loading is usually centered on the footing. If a single footing interferes with another spread footing, the two can be combined to form a combined footing (Figure 4.1(b)). For foundations supporting two or more column loads, or machinery, the positioning of the loading or weight often makes a rectangular or trapezoidal shape necessary.

Mat (or raft) foundations

The mat (or raft) foundation can be considered a large footing extending over a great area, frequently an entire building area (Figure 4.1 (d)). All vertical structural loadings from columns and walls are supported on the common foundation. Typically, the mat is utilized for conditions where a preliminary design indicates individual column footing would undesirably close together or try to overlap. The mat is frequently utilized as a method to reduce or distribute building loads in order to reduce differential settlement between adjacent areas. To function properly, the mat structure will be more rigid and thicker than the individual spread footing. Some rafts are thickened in the form of a waffle. If the overburden pressure of a soil is removed to embed a raft foundation is equal to the stress imposed by the structure, the raft is termed fully compensated or floating raft. If the overburden pressure is a fraction of imposed stress, the raft is partially compensated.

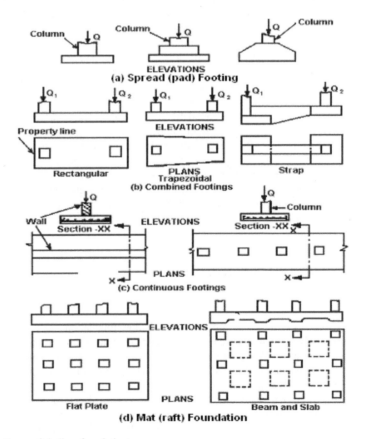

Figure 4.1.   Types of shallow foundations.

4.3   BEARING CAPACITY

Bearing capacity is a general term used to describe, the load carrying capacity of a foundation soil or rock which enables it to bear and transmit loads from a structure. The ultimate soil bearing capacity for foundations (the loading that will cause a shear failure in the supporting soils) is related to the properties of the soil, including the past stress history and the proximity of the groundwater table; it is also affected by the characteristics of the foundation, including size, depth, shape, and the method of construction or installation.

The three principal modes of soil failure, established by the patterns of the shearing soil zones, are defined as general shear failure, local shear failure, and punching shear failure. Figure 4.2 illustrates the differences in the three modes and the foundation load settlement curve typical to each mode.

The general shear failure (Figure 4.2a), expected for soils possessing brittle-type stress–strain characteristics, is identified by a well-defined wedge beneath the foundation and slip surfaces extending diagonally from the side edges of the footing adjacent to the footing bulges upward to the ground surface. Soil displacement is accompanied by tilting of the foundation(unless the foundation is restrained). The load-displacement curve for the general shear case indicates that failure is abrupt.

The punching shear failure (Figure 4.2c), occurring in soil possessing the stress–strain characteristics of a very plastic material, involves poorly defined shearing planes. Significant compression of a wedge-shaped soil zone beneath the foundation is accompanied by the occurrence of vertical shear beneath the edges of the foundation. The soil zones beyond the edges of the foundation are little affected, and no significant degree of surface bulging occurs. Aside from large settlement, failure is not clearly recognized.

The local shear failure condition (Figure 4.2b) involves some of the characteristics of both the general shear and the punching shear failure modes. As for general shear, the well-defined wedge and slip surfaces are formed beneath the foundation, but the slip surfaces fade into the soil mass

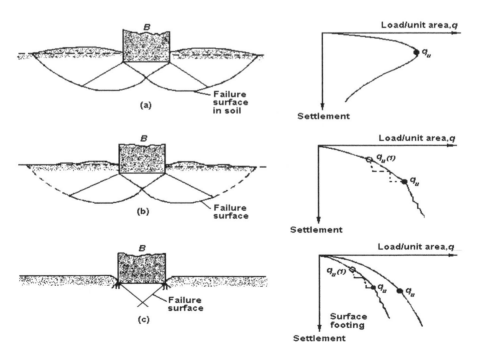

Figure 4.2.   Bearing capacity failure.

beyond the edges of the foundation and do not carry upward to the ground surface. Slight bulging of the ground surface adjacent to the foundation does occur. As with punching shear, significant vertical compression of soil directly beneath the foundation takes place. The local shear condition represents a transitional mode between the general shear and the punching shear failures and is expected for soils possessing somewhat plastic stress-strain characteristics.

The load-settlement curves of Figure 4.2 are qualitative in nature. Investigators of bearing capacity problem have found that relatively large settlements are required in order to have foundations reach the "failure" load (viz., 3 to 15% of the foundation width for very shallow installations and up to 25% of the foundation width for deeper installations).

For application to shallow foundation design, it is commonly considered that the general shear case applies to dense granular soil and to firmer saturated cohesive soils subject to undrained loading. The punching shear case is considered appropriate for compressible soil, such as sands having a low to medium relative density, and for cohesive soils subject to slow loading.

Methods in widespread use for determining soil bearing capacity include the application of bearing capacity equations, the utilization of penetration resistance data obtained during soil explorations, and the practice of relating the soil type to a presumptive bearing capacity recommended by building codes. Permissible bearing capacities determined by the equation method or from building code tables typically do not consider effects of soil compressibility and the possible influence of poorer soil layers underlying the bearing layer. Consequently, settlement determinations and other results of soil deformation must be analyzed separately. Foundation design criteria developed from boring-penetration resistance data often relate a foundation bearing pressure to settlement. Design data in this form are convenient to use, but the methods available do not cover all foundations and soil types. An additional method in use for determining the permissible or safe design loading is the field load test performed on a in-place foundation unit. Load tests relate carrying capacity and settlement together, which is an advantage. Disadvantages include the cost and time involved. Further, test results require care in their evaluation, for it is known that a load test on a small shallow foundation may not be representative of the behavior of a large shallow foundation.

Of the common procedures for foundation design, the analytical method using soil mechanics principles (e.g., the use of bearing capacity equations with settlement analysis) and the use of penetration resistance data are preferred. Properly applied, these methods consider the effects of foundation type and size as well as the properties of soil to the depth that will have a significant effect on the foundation performance. Load test data will provide reliable design information if properly related to the results of subsurface exploration and a final foundation design. The use of presumptive bearing capacities is discouraged because of the heavy reliance on soil-type description, with little correlation to the soils' actual physical properties and no consideration of the possible existence of poor soil strata underlying the foundation bearing level.

Important definitions in bearing capacity analyses

The following definitions are used in bearing capacity analyses.

a. *Foundation soil or bed* is the soil to which loads are transmitted from the base of the structure.
b. *Footing* is an enlarged base of the structure to distribute the column or wall load to the ground at a compatible strength and deformation characteristics of foundation soil.
c. *Mat or raft* is characterized by the feature that columns or wall frames are set into the footing in two directions. Any number of columns can be accommodated with as low as four columns.
d. *Gross bearing pressure*, q, is total load divided by area of footing.
e. *Net bearing pressure*, q_n is net increase in pressure at level of footing, that is the gross bearing capacity minus the original overburden pressure or surcharge pressure at the foundation level. This however will be the same as the gross bearing pressure when the depth of foundation is zero (foundation at ground surface).

f. *Ultimate bearing capacity*, q_f, is earth pressure at shear failure, that is the maximum pressure which a foundation soil can withstand till the onset of occurrence of shear failure of the foundation.

g. *Net ultimate bearing pressure*, q_{fn}, is the difference between ultimate bearing capacity with pressure due to weight of soil ($q_f - \gamma z$), where γ is soil unit weight and z is depth of footing base.

h. *Maximum safe bearing capacity*, q_s is intensity of load that can be supported by the soil without risk of shear failure (that is ultimate bearing capacity divided by a suitable factor of safety). The factor of safety in foundations may range from 2 to 5, depending upon the importance of the structure, and the soil profile at the site. The factor of safety should be applied to the net ultimate bearing capacity, and the surcharge pressure should then be added to get the safe bearing capacity. It is thus the maximum intensity of loading which can be transmitted to the soil without the risk of shear failure, irrespective of the settlement that may occur.

i. *Allowable bearing pressure*, q_a is the maximum allowable net loading intensity on the soil at which the soil neither fails in shear nor undergoes excessive or intolerable settlement detrimental to the structure.

Factors affecting bearing capacity

Bearing capacity is governed by a number of factors. The following are the some of the more important factors that affect bearing capacity.

i. Nature of soil and its physical and engineering properties.
ii. Nature of the foundation and other details such as the size, shape, depth at which the foundation is located and rigidity of the structure.
iii. Total and differential settlement that the structure can withstand without functional failure.
iv. Location of ground water table relative to the level of foundation.
v. Initial stresses, if any.

In view of the wide variety of factors that affect the bearing capacity, a systematic study of the factors involved in a logical sequence is necessary for a proper understanding.

Bearing capacity equations

Historically, a number of investigators have undertaken studies relating to foundation bearing capacity, typically applying the classical theories of elasticity and plasticity to soil behavior to develop equations appropriate for foundation design. Behavior described by the classical theory of plasticity has been widely used to obtain a solution for the case of a general shear failure. The original theoretical concepts for analyzing conditions considered applicable to foundation performance using the theory of plasticity are credited to Prandtl (1920) and Reissner (1924). Prandtl studied the effect of a long, narrow metal tool bearing against the surface of a smooth metal mass that possessed cohesion and internal friction but no weight. The results of Prandtl's work were extended by Reissner to include the condition where the bearing area is located below the surface of the resisting material and a surcharge weight acts on a plane that is level with the bearing area. Terzaghi (1943) applied the developments of Prandtl and Reissner to soil foundation problems, extending the theory to consider rough foundation surfaces bearing on materials that possess weight. Terzaghi developed general bearing capacity equations for a strip footing that combined the effects of soil cohesion and internal friction, foundation size, soil weight and surcharge effects in order to simplify the calculations necessary for foundation design. His equations utilized the concept of a dimensionless bearing capacity factor whose values are a function of the shear possessed by the supporting soil. Through the ensuing years, the ultimate bearing capacity for shallow and deep foundations has continued to be studied in the quest for a refined definition of foundation-soil behavior and a generalized bearing capacity equation which agrees well with failure conditions occurring in model and large scale foundations (Meyerhof (1951), Hansen (1970), Vesic (1975) and De Beer (1970). Modifications to early concepts have emerged from such studies, but the general form of the Terzaghi bearing capacity equations has been retained because of its practicality.

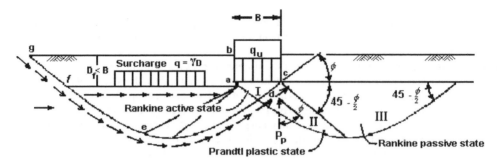

Figure 4.3.   Terzaghi failure mechanism.

Terzaghi's bearing capacity theory
The assumed failure mechanism of Terzaghi is illustrated in Figure 4.3. Terzaghi assumed a strip footing with rough base placed at a depth D_f on a homogeneous and isotropic soil mass. In the analysis, the shearing resistance of the soil above the base (*ab* in Figure 4.3) of the footing is neglected, but the effect of soil weight above the base is considered by superimposing an equivalent surcharge of intensity $q = \gamma D_f$. The development of the failure surface in the soil is governed by the general shear failure.

The soil immediately beneath the foundation forms a wedge (zone I) which moves downwards. The movement of the wedge forces the soil aside and produces two zones of shear (zone II and zone III), consisting of a radial shear zone (zone II) and a linear shear zone (zone III). Zone I is considered to be at Rankine active state, zone II under radial shear and zone III at Rankine passive state. At the verge of failure, $\Sigma V = 0$, thus

$$q_{ult}B = 2P_p + 2\,ad\,c\sin\phi$$

Substituting $ad = B/2 \cdot cos\,\phi$,

$$q_{ult}B = 2P_p + Bc\tan\phi$$

The value of P_p has been represented as the vector sum of three components, (i) cohesion, (ii) surcharge and (iii) weight of the soil. Terzaghi assumed the method of superposition to be valid and presented the unit ultimate bearing capacity in the form,

$$q_{ult} = cN_c + \tfrac{1}{2}B\gamma N\gamma + qN_q \tag{4.1}$$

where, N_c, N_γ and N_q are non-dimensional bearing capacity factors and function only of the angle of shearing resistance, ϕ.

$$N_c = \cot\phi[2(3\pi/4 - \phi/2)\tan\phi]/2\cos^2(\pi/4 + \phi/2) \tag{4.2a}$$

$$N_q = \exp[N_q - 1] \tag{4.2b}$$

$$N_\gamma = \tan\varphi(k_{p_\gamma}/\cos^2\varphi - 1) \tag{4.2c}$$

A close approximation of $k_{p\gamma}$ is given by

$$k_{p\gamma} = 3\tan^2[45 + (\varphi + 33)/2]$$

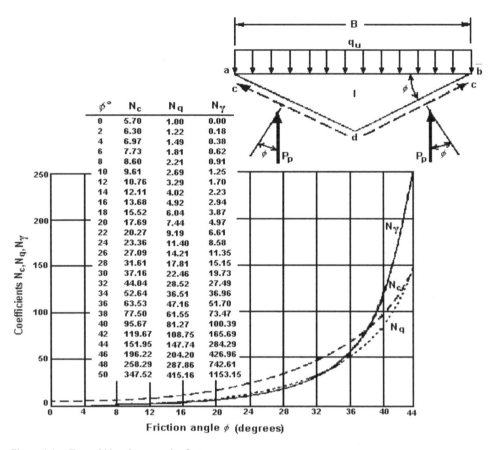

Figure 4.4.   Terzaghi bearing capacity factors.

Terzaghi bearing capacity factors are shown in Figure 4.4. With the addition for shape factors in the cohesion and base terms, Terzaghi obtained the following expressions for the ultimate bearing capacity for general shear conditions.

Long footings: $q_{ult} = cN_c + \frac{1}{2}B\gamma N\gamma + qN_q$

Square footing: $q_{ult} = 1.3\,cN_c + 0.4B\gamma N\gamma + qN_q$

Circular footings: $q_{ult} = 1.3\,cN_c + 0.3B\gamma N\gamma + qN_q$

Terzaghi developed his bearing capacity equations assuming a general shear failure.

Meyerhof's bearing capacity theory
Meyerhof (1951) considered the effects of shearing resistance within the soil above foundation level, the shape and roughness of foundation, and derived a general bearing capacity equation. His expression for the shape, depth and inclination factors, and the bearing capacity factors are shown in Table 4.1 and Figure 4.5. According to Meyerhof,

for vertical load: $q_{ult} = cN_cS_cd_c + \frac{1}{2}B\gamma N\gamma S\gamma d\gamma + qN_qS_qd_q$

for inclined load: $q_{ult} = cN_cS_cd_ci_c + \frac{1}{2}B\gamma N\gamma S\gamma d\gamma i\gamma + qN_qS_qd_qi_q$

Table 4.1.   Meyerhof bearing capacity factors.

For ϕ	Shape	Depth	Inclination
Any	$s_c = 1 + 0.2 k_p \dfrac{B}{L}$	$d_c = 1 + 0.2 \sqrt{k_p} \dfrac{D}{B}$	$i_c = i_q = \left(1 - \dfrac{\alpha}{90°}\right)^2$
For $\phi = 0°$	$s_q = s_\gamma = 1.0$	$d_q = d_\gamma = 1.0$	$i_\gamma = 1$
For $\phi \geq 10°$	$s_q = s_\gamma = 1 + 0.1 k_p \dfrac{B}{L}$	$d_q = d_\gamma = 1 + 0.1 \sqrt{k_p} \dfrac{D}{B}$	$i_\gamma = \left(1 - \dfrac{\alpha}{90°}\right)^2$

$k_p = \tan^2 \left(45 + \dfrac{\phi}{2}\right)$

α = angle of resultant measured from vertical axis

When triaxial ϕ is used for plane strain, adjust ϕ to

ϕ_{triaxial} obtain $\phi_{ps} = \left(1.1 - 0.1 \dfrac{B}{L}\right)$

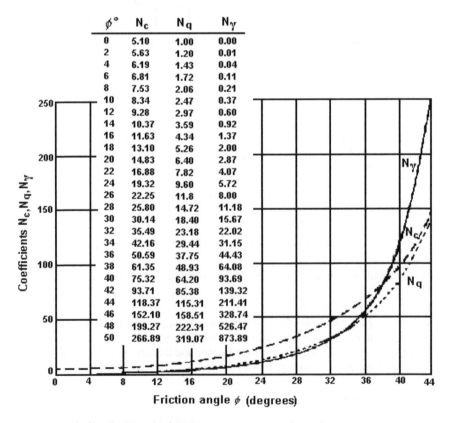

$\phi°$	N_c	N_q	N_γ
0	5.10	1.00	0.00
2	5.63	1.20	0.01
4	6.19	1.43	0.04
6	6.81	1.72	0.11
8	7.53	2.06	0.21
10	8.34	2.47	0.37
12	9.28	2.97	0.60
14	10.37	3.59	0.92
16	11.63	4.34	1.37
18	13.10	5.26	2.00
20	14.83	6.40	2.87
22	16.88	7.82	4.07
24	19.32	9.60	5.72
26	22.25	11.8	8.00
28	25.80	14.72	11.18
30	30.14	18.40	15.67
32	35.49	23.18	22.02
34	42.16	29.44	31.15
36	50.59	37.75	44.43
38	61.35	48.93	64.08
40	75.32	64.20	93.69
42	93.71	85.38	139.32
44	118.37	115.31	211.41
46	152.10	158.51	328.74
48	199.27	222.31	526.47
50	266.89	319.07	873.89

Figure 4.5.   Meyerhof bearing capacity factors.

$$\text{where} \quad N_q = \exp^{\pi \tan \phi} \tan^2 (\pi/4 + \varphi/2) \tag{4.3a}$$

$$N_c = (N_q - 1) \cot \varphi \tag{4.3b}$$

$$N_\gamma = (N_q - 1) \tan (1.4 \, \varphi) \tag{4.3c}$$

For footing subjected to eccentric loading, Meyerhof suggested the following correction:

$$B' = B - 2e \tag{4.4}$$

where B' = reduced width of footing to account for effect of load with eccentricity, e; B = original width of footing and e = eccentricity measured from the symmetrical axis of the footing.

Next, vertical footing pressure, q, can be calculated as follows:

$$q = P/(B' \times L) \tag{4.5}$$

where, q = vertical footing pressure; P = vertical load and L = length of footing.

Hansen's bearing capacity theory

Hansen (1970) proposed a more generalized equation with shape and depth of foundation and the inclination of the load.

$$q_{ult} = cN_cS_cd_ci_cb_cg_c + \tfrac{1}{2}B_\gamma N_\gamma S_\gamma d_\gamma i_\gamma b_\gamma g_\gamma + qN_qS_qd_qi_qb_qg_q \tag{4.6}$$

Meyerhof and Hansen's theories give almost identical N_c and N_q values. The N_γ value recommended by Hansen is almost the same as Meyerhof's for φ values up to 35°. There are some deviations for higher values (Table 4.2). Figure 4.6 illustrates Hansen's bearing capacity factors. A comparative statement of the bearing capacity factors is presented in Table 4.3.

For the case of an inclined load, Hansen's solution for equivalent vertical pressure, q_e, can be simplified as follows:

$$q_e = (P_v + \lambda P_h)/A \tag{4.7}$$

where, A = base area of footing ($B' \times L$); P_v = component of vertical load; P_h = component of horizontal load; λ = empirical dimensionless constant depending on angle of friction, ϕ (see Table 4.4).

Skempton's theory

Skempton (1951) proposed equations for bearing capacity of footings founded on purely cohesive soils based on extensive investigations. According to Skempton, the bearing capacity factors N_c is a function of the depth of foundation and also of its shape. The equation for net ultimate bearing capacity, $q_{net \, ult}$ is as follows:

$$q_{net \, ult} = cN_c \tag{4.8}$$

The bearing capacity factor, N_c is given by the following:

$$\text{Strip footing: } N_c = 5(1 + 0.2D_f/B) \tag{4.9}$$

(with a limiting value of N_c of 7.5 for D_f/B greater than 2.5).

$$\text{Square and circular footings: } N_c = 6(1 + 0.2D_f/B) \tag{4.10}$$

(with a limiting value of N_c of 9.0 for D_f/B greater than 2.5. B is the width of strip, side of square or diameter of a circular footing).

$$\text{Rectangular footings: } N_c = 5(1 + 0.2B/L)(1 + 0.2D_f/B) \tag{4.11}$$

Table 4.2.   Hansen's bearing capacity factors.

Shape Factors

Since failure can take place either along the long sides, or along the short sides, Brinch Hansen proposed two sets of shape factors.

$$s_{cB}^a = 0.2\, i_{cB}^a\, BL$$
$$s_d^a = 0.2\, i_d^a\, LB$$
$$s_{qB} = 1 + \sin\phi.Bi_{qB}/L$$
$$s_{qL} = 1 + \sin\phi.Li_{qL}/B$$
$$s_{\gamma B} = 1 - 0.4(Bi_{\gamma B}) : (Li_{\gamma L})$$
$$s_{\gamma L} = 1 - 0.4(Li_{\gamma L}) : (Bi_{\gamma B})$$

For the last two factors, the special rule must be followed, that the value exceeding 0.6 should always be used.

Load Inclination Factors

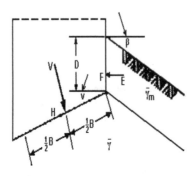

Base and Ground Inclination Factors

$$b_r^a = \frac{2v}{\pi+2} = \frac{v^0}{147^0} \quad b_q = e^{-2p\tan\phi} \quad b_\gamma = e^{-2.7v\tan\phi}$$
$$g_r^a = \frac{2\beta}{\pi+2} = \frac{\beta^0}{147^0} \quad g_q = [1 - 0.5\tan\beta]^5 = g_\lambda$$

$$i_c^a = 0.5 - 0.5\sqrt{1 - H/Ac_w}$$
$$i_q = [1 - 0.5H : (V + Ac\cot\phi)]^5$$
$$\text{for } v = 0^0$$
$$i_\gamma = [1 - 0.7H : (V + Ac\cot\phi)]^5$$
$$i_\gamma = [1 - (0.7 - v^0/450^0)H : (V + Ac\cot\phi)]^5 \quad v > 0^0$$

Depth Factors

$$d_c^a = 0.4\, D/B$$
$$\text{for } D \leqq B$$
$$d_q = 1 + 2\tan\phi(1 - \sin\phi)^2 D/B$$
$$d_\gamma^a = 0.4\, arc\tan D/B$$
$$\text{for } D > B$$
$$d_q = 1 + 2\tan\phi(1 - \sin\phi)^2 arc\tan D/B$$
$$d_\gamma = 1$$

(for $D_f/B \leq 2.5$), and

$$N_c = 7.5(1 + 0.2B/L) \tag{4.12}$$

(for $D_f/B > 2.5$)

where, B and L are breadth and length respectively, of the rectangular footing.

4.4   FOUNDATION DESIGN USING PENETRATION RESISTANCE DATA

Obtaining foundation design criteria directly from the soil borings is normally performed to explore if conditions underlying a site are desirable from the view of cost and time savings. At locations where it is difficult to recover soil samples suitable for determining soil properties (a requirement when an analytical study for a foundation design is to be performed), obtaining foundation design criteria based directly on exploration data may be necessary.

Empirically developed foundation design methods are typically based on the averaged or representative performance of case studies, but they will be conservative. Practically, deviations will always exist between individual cases and the design recommendations.

Often, empirical design methods are recommended only for preliminary studies. Realistically, designers frequently need also to apply the procedures to final designs because of lack of further information.

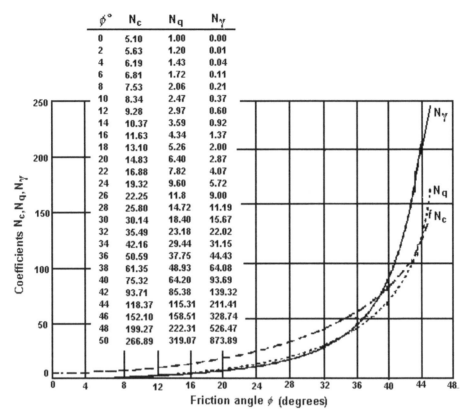

$\phi°$	N_c	N_q	N_γ
0	5.10	1.00	0.00
2	5.63	1.20	0.01
4	6.19	1.43	0.04
6	6.81	1.72	0.11
8	7.53	2.06	0.21
10	8.34	2.47	0.37
12	9.28	2.97	0.60
14	10.37	3.59	0.92
16	11.63	4.34	1.37
18	13.10	5.26	2.00
20	14.83	6.40	2.87
22	16.88	7.82	4.07
24	19.32	9.60	5.72
26	22.25	11.8	9.00
28	25.80	14.72	11.19
30	30.14	18.40	15.67
32	35.49	23.18	22.02
34	42.16	29.44	31.15
36	50.59	37.75	44.43
38	61.35	48.93	64.08
40	75.32	64.20	93.69
42	93.71	85.38	139.32
44	118.37	115.31	211.41
46	152.10	158.51	328.74
48	199.27	222.31	526.47
50	266.89	319.07	873.89

Figure 4.6.   Hansen bearing capacity factors.

Standard penetration test – cohesionless soil

Early information relating to standard penetration test results in sand to spread footing size, bearing pressure, and settlement was presented by Terzaghi and Peck in 1968 and has been widely referred to for foundation design. This empirically determined design method was intended to provide foundations whose maximum settlement would not exceed one inch, with the expectation that the greatest differential settlement between different footings would not exceed one-half to three-fourths inch, a tolerable range for most structures. These early recommendations were based on limited data and were deliberately conservative.

Subsequent studies, which included performance evaluations of foundations designed in accord with the original recommendations, have indicated that modification of the original design criteria is necessary where greater allowable bearing pressures can be permitted. When SPT values are used for foundation design, the field blow counts need to be corrected to reflect the effect of sample depth on actual soil properties.

Figure 4.7 presents design criteria for spread footings located at a shallow installation depth. The bearing pressures are intended to be values that produce a maximum settlement to the order of one inch. The maximum differential settlement that results between all foundation units on the site is expected to be less than three-fourths of an inch. The N values used for design are corrected values ($N_{corr} = C_N N_{field}$; $C_N = \sqrt{1/\sigma_v}$; see Figure 4.8). The design bearing pressure shown in Figure 4.7 presume the water table to be greater than at depth B (foundation width) below the bottom of the footing. If the water table is within the zone close to the bottom of the footing, a bearing pressure value obtained from the design curves should be reduced to keep settlement from exceeding the one-inch limit. For the condition where the water table is at the base of the foundation, a one-third

Table 4.3.   A comparative statement of bearing capacity factor.

ϕ^0	Terzaghi			Meyerhof			Hansen		
	N_c	N_q	N_γ	N_c	N_q	N_γ	N_c	N_q	N_γ
0	5.70	1.00	0.00	5.10	1.00	0.00	5.10	1.00	0.00
2	6.30	1.22	0.18	5.63	1.20	0.01	5.63	1.20	0.01
4	6.97	1.49	0.38	6.19	1.43	0.04	6.19	1.43	0.05
6	7.73	1.81	0.62	6.81	1.72	0.11	6.81	1.72	0.11
8	8.60	2.21	0.91	7.53	2.06	0.21	7.53	2.06	0.22
10	9.61	2.69	1.25	8.34	2.47	0.37	8.34	2.47	0.39
12	10.76	3.29	1.70	9.28	2.97	0.60	9.28	2.97	0.63
14	12.11	4.02	2.23	10.37	3.59	0.92	10.37	3.59	0.97
16	13.68	4.92	2.94	11.63	4.34	1.37	11.63	4.34	1.43
18	15.52	6.04	3.87	13.10	5.26	2.00	13.10	5.26	2.08
20	17.69	7.44	4.97	14.83	6.40	2.87	14.83	6.40	2.95
22	20.27	9.19	6.61	16.88	7.82	4.07	16.88	7.82	4.13
24	23.36	11.40	8.58	19.32	9.60	5.72	19.32	9.60	5.75
26	27.09	14.21	11.35	22.25	11.85	8.00	22.25	11.85	7.94
28	31.61	17.81	15.15	25.80	14.72	11.19	25.80	14.72	10.94
30	37.16	22.46	19.73	30.14	18.40	15.67	30.14	18.40	15.07
32	44.04	28.52	27.49	35.49	23.18	22.02	35.49	23.18	20.79
34	52.64	36.51	36.96	42.16	29.44	31.15	42.16	29.44	28.77
36	63.53	47.16	51.70	50.59	37.75	44.43	50.59	37.75	40.05
38	77.50	61.55	73.47	61.35	48.93	64.08	61.35	48.93	56.18
40	95.67	81.27	100.39	75.32	64.20	93.69	75.32	64.20	79.54
42	119.67	108.75	165.69	93.71	85.38	139.32	93.71	85.38	113.96
44	151.95	147.74	248.29	118.37	115.31	211.41.	118.37	115.31	165.58
46	196.22	204.20	426.96	152.10	158.51	329.74	152.10	158.51	244.65
48	258.29	287.86	742.61	199.27	222.31	526.47	199.27	222.31	368.68
50	347.52	415.16	1153.15	266.89	319.07	873.89	266.89	319.07	568.59

Table 4.4.   Factors for calculating bearing capacity of inclined load.

$\tan \phi/F$	0	0.2	0.4	0.6	0.8	1.0
λ		1.4	1.8	2.3	2.8	3.3	3.9

reduction in the bearing pressure value should be applied. A linear interpolation can be assumed for water table depths intermediate between the formation level and a distance B below it.

Standard penetration test – cohesive soil

Shallow foundations supported on clay are not designed directly from standard penetration test blow count data. A major reason for this situation has been the difficulty in relating blow count values to a cohesive soil stress history accurately enough to permit reliable foundation settlement predictions. Since practical procedures exist for separately determining the properties of a clay (vane shear tests can be performed by boring on in-place soil, or laboratory strength and compressibility tests can be performed on undisturbed samples recovered from borings), some of the usual reasons for converting boring results directly into foundation design data do not apply. However, blow counts from the standard penetration test are useful for estimating the value of cohesion, c:

$$c = N/4 \text{ ksf (for clay)}$$

$$c = N/5 \text{ ksf (for silty clay)}$$

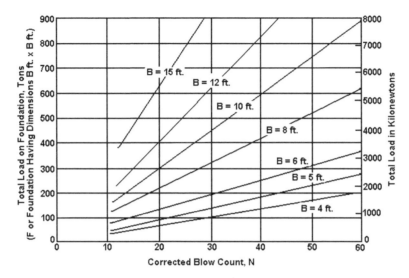

Figure 4.7. Chart for spread footings on sand. Corrected standard penetration test results related to foundation size and loading for approximately one-inch settlement. Water table at more than depth B below footing.

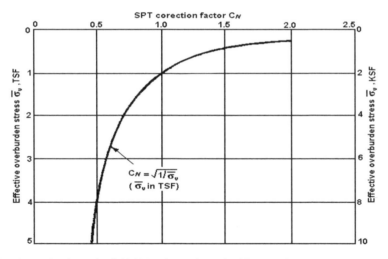

Figure 4.8. Correction factor for field SPT values (Liao and Whitman 1986).

In Malaysia, light weight dynamic penetrometer (a.k.a JKR dynamic cone penetrometer) is also widely used for preliminary assessment of subsoil conditions. Figure 4.9 shows an example of a design chart for allowable bearing capacity based on the penetrometer value.

Static cone resistance – cohesionless soil

In recent times, investigations utilizing penetrometer methods have received considerable attention in virtually all areas of the world. Interest has been generated by developing awareness that direct relationships between penetrometer results and foundation design have been formulated.

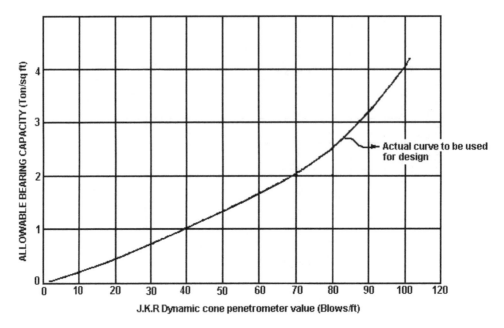

Figure 4.9.   Allowable bearing capacity.

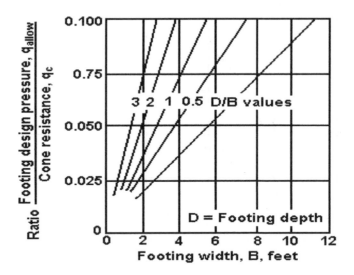

Figure 4.10.   Relationship among static cone resistance, footing size and length, and allowable bearing pressure (Sanglerat 1972).

An example of design information, reflecting the effect of footing width and depth is shown in Figure 4.10.

Figure 4.11 presents design criteria for shallow footings with the bearing pressures intended to be values that produce a maximum settlement to the order of one inch. It is assumed that the value of point resistance, q_c is representative of the soil zone below the base of the footing to a depth extending approximately $1.5B$ further, and the water table is deep.

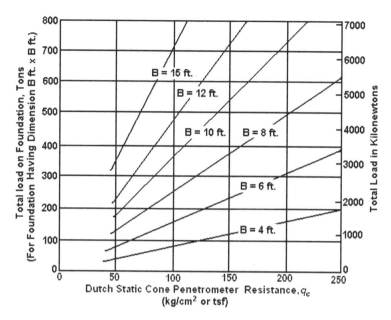

Figure 4.11.   Chart for spread footings on sand. Dutch static cone resistance related to foundation size and loading for approximately one-inch settlement. Water table at more than depth B below footing.

Static cone resistance – in cohesive soil

In practical terms, good relationships have been established between q_c values and the cohesion and stress history of clay. The bearing capacity for the clay is subsequently calculated from the theoretical bearing capacity equation. For Dutch Cone:

$c =$ between $q_c/15$ and $q_c/18$ (for normally consolidated clay)
$c =$ between $q_c/22$ and $q_c/26$ (for overconsolidated clay)

4.5   PRESSUREMETER TEST

The ultimate bearing capacity, q_{ult}, for the soil zone supporting a shallow foundation, based on theoretical considerations with empirically determined modifications, has been related to the limit pressure as follows:

$$q_{ult} = \sigma_v + k_{bc}(p_l - \sigma_h)$$

where, $q_{ult} =$ ultimate bearing capacity of the soil, in kN/m²; $\sigma_v =$ effective vertical stress at the planned foundation depth, in kN/m²; $k_{bc} =$ bearing capacity factor for the proposed foundation shape and embedment and soil type (see Table 4.5); $p_l =$ limit pressure; $\sigma_h =$ at-rest horizontal soil pressure at the planned foundation depth in kN/m².

The settlement, ΔH, of a shallow foundation situated in homogeneous soil can be estimated from:

$$\Delta H = q_{des}/9E_m[2\,B_o(\lambda_d B/B_o)^\alpha + \alpha\lambda_c B]F_d$$

Where, $q_{des} =$ foundation design bearing pressure minus the vertical pressure of the soil overburden adjacent to the foundation base; $B_o =$ a reference dimension, equal to 0.6 m or 2 ft, $B =$ width or diameter of the foundation, provided that B is equal to or greater than B_o; $\alpha =$ rheological or creep deformation factor, which depends on the soil type and the ratio (Table 4.6); $E_m/(P_l - \sigma_h)$; $\lambda_c, \lambda_d =$ shape factors that are based on length-to-width ratio of the foundation (Table 4.7);

Table 4.5.   Values of k_{bc} for pressuremeter bearing capacity equation.

Soil type:	Sand			Silt			Clay		
D/B ratio:	0	2	4	0	2	4	0	2	4
P_t (kN/m^2)									
Values of k_{bc}, strip foundations									
500	0.8	1.5	1.8	0.8	1.4	1.6	0.8	1.3	1.5
1000	0.8	1.7	2.2	0.8	1.45	1.75	0.8	1.4	1.7
3000	0.8	2.0	2.6	0.8	1.65	2.0	0.8	1.5	1.8
6000	0.8	2.3	3.0	–	–	–	–	–	–
Values of k_{bc}, square and round foundations									
500	0.8	2.2	3.0	0.8	2.1	2.45	0.8	1.9	–
1000	0.8	3.0	4.0	0.8	2.25	2.75	0.8	2.2	2.5
3000	0.8	3.7	4.8	0.8	2.45	3.1	0.8	2.4	2.9
6000	0.8	3.9	5.4	–	–	–	–	–	–

Table 4.6.   Pressuremeter settlement equation, α factors.

Soil type	$E_m/(p_t - \sigma_h)$	α
Clay, normally consolidated	9 to 16	2/3
Clay, overconsolidated	Greater than 16	1
Silt, normally consolidated	8 to 14	1/2
Silt, overconsolidated	Greater than 14	2/3
Sand	7 to 12	1/3
Sand, compact	Greater than 12	1/2

Table 4.7.   Pressuremeter settlement equation, λ shape factors.

L/B ratio	Circular	Square	2	5	20
λ_c	1	1.10	1.20	1.40	1.50
λ_d	1	1.12	1.53	2.14	2.65

E_m = Menard pressuremeter modulus, equal to $2.66G_m$, where G_m is the Menard shear modulus for the pressuremeter curve between, v_o and v_r, i.e. $G_m = \{V_c + (v_o + v_r)/2\}((p_f - p_o)/(v_f - v_o))$, where the respective pressure and volume values are obtained from the pressuremeter curve (see Figure 4.12); F_d = depth factor, equal to 1 if foundation depth is greater than B, equal to 1.2 for a foundation at the ground surface, equal to 1.1 for a foundation depth equal to one-half B.

4.6   BEARING CAPACITY FROM FIELD LOAD TESTS

Obviously the most reliable method of obtaining the ultimate bearing capacity at a site is to perform a load test. This would directly give the bearing capacity if the load test is on a full size footing; however, this is not usually done since an enormous load would have to be applied. Another factor is that the bearing capacity obtained is for that size and if there is more than one size then additional tests would be required. For the test just described, the cost could be very high.

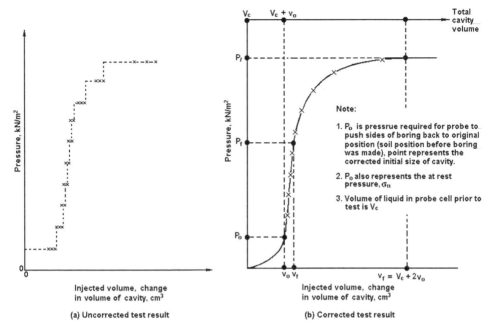

Figure 4.12. Pressuremeter test results.

The usual practice is to load test small plates of diameters from 20 to 30 inches or square of side 12×12 and perhaps 24×24 inches. These sizes are usually too small to extrapolate to full size footings which may be 1.5 to 5 meters square. Some of the factors causing the extrapolation to be questionable are:

a. The influence depth is significantly different for the model versus prototype footing. Any stratification below the influence depth has minimal effect on the model but may be a major influence on the full size footing.
b. The soil at greater depths has more overburden pressure acting to confine the soil so it is effectively "stiffer" than the near surface soils. This markedly affects the load-settlement response which is used to define q_{ult}.

Extrapolating load test results to full size footings is not standard. It is common for clay soils to note that $BN\gamma$ term is zero so that one might say that q_{ult} is independent of footing size giving

$$q_{\text{ult,foundation}} = q_{\text{ult,load test}}$$

In cohesionless soils, all three terms of the bearing capacity equation applies and noting the N_γ term includes the footing width, one might say

$$q_{\text{ult,foundation}} = M + N(B_{\text{foundation}})/B_{\text{load test}}$$

where M includes the N_c and N_q terms and N is the N_γ term. By using several size plates this equation can be solved graphically. Practically, for extrapolating plate load tests which are often in a configuration so that the N_q term is negligible, and for sands, is to use

$$q_{\text{ult}} = q_{\text{plate}}(B_{\text{foundation}})/B_{\text{plate}}$$

The use of this equation is not recommended unless the $(B_{\text{foundation}})/B_{\text{plate}}$ is not much less than about 3. When the ratio is about 6 to 15 or more, the extrapolation from a plate load test is little more than a guess that could be obtained at least as reliably as using an SPT or CPT correlation.

Table 4.8.   Presumptive unit soil bearing values.

Class	Material	Allowable bearing value, tons per square foot[1]
1	Massive crystalline bed rocks, such as granite, gneiss, trap rock, etc.; in sound condition	100
2	Foliated rocks, such as schist and slate, in sound condition	40
3	Sedimentary rocks, such as hard shales, silt stones, or sandstones, in sound condition	15
4	Exceptionally compacted gravels or sands	10
5	Gravel; sand-gravel mixtures; compact	6
6	Gravel, loose; coarse sand, compact	4
7	Coarse sand, loose; sand-gravel mixtures, loose; fine sand, compact; coarse sand, wet (confined)	3
8	Fine sand, loose; fine sand, wet (confined)	2
9	Stiff clay	4
10	Medium stiff clay	2
11	Soft clay	1
12	Fill, organic material, or silt	([2])

[1] Presumptive bearing values apply to loading at the surface or where permanent lateral support for the bearing soil is not provided.
[2] Except where, in the opinion of the enforcement officer, the bearing value is adequate for light frame structures, fill material, organic material, and silt shall be deemed to be without presumptive bearing value. The bearing value of such material may be fixed on the basis of tests or other satisfactory evidence.

4.7   PRESUMPTIVE BEARING PRESSURES

Allowable foundation bearing values that are related to the visual classification of a site's bearing soil, a type of design information that conventionally appears as part of the building codes, are termed presumptive bearing pressures. Examples are shown in Table 4.8.

Presumptive bearing pressures can be appropriate for design when utilized in the area for which the values were developed and for the type of construction that provided the original experience data. Unfortunately, presumptive bearing pressure frequently no longer reflects local geologic conditions and experiences but are values extracted from other codes or references. Importantly, methods and materials of construction that affect the behavior of commercial-industrial buildings are likely have changed significantly since the origin of the presumptive bearing values for places such as older cities; for example, compared to the older style of low masonry construction. Modern structures are taller, with steel frame construction, and the performance demands of foundations have changed correspondingly. Since presumptive bearing pressures do not consider the important influence of factors such as foundation size and embedment, position of the ground water table, soil density or consistency, and soil stress history, unsafe foundation design can occur. It is emphasized that the presumptive bearing pressures should be used by the engineer only for preliminary foundation design purposes and, in all cases, he should then review and, if necessary, amend his first design.

4.8   SETTLEMENT

Settlements produced by isolated individual footing are due to two sources viz:

(i) deep-seated consolidation from volume change, and
(ii) immediate settlement from shear beneath the individual footings.

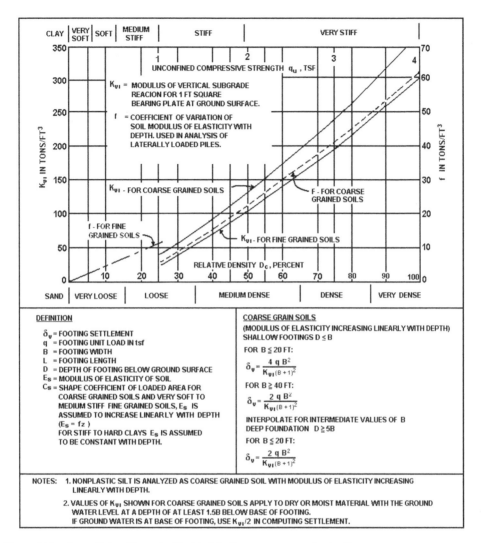

Figure 4.13.   Immediate settlement of isolated footings on course grained soils.

Consolidation settlement

The term consolidation denotes a shallow process of compression due to extrusion of water from the voids in the fine-grained soils.

Extrusion of water from cohesionless soils causes settlement to take place almost immediately as the foundation loading is imposed.

The procedure for estimating consolidation settlements is conventionally based on Skempton and Bjerrum's modification of Terzaghi's theory of consolidation. The *p–e* curve is normally obtained from an oedometer test on a saturated sample.

However, not all foundation soils are saturated especially residual soils. Consequently the oedometer test will not yield a practical estimate of settlement for our residual soils.

Shear strain settlement

Where immediate settlements dominate, compute values by the procedure of Figure 4.13 for coarse grained soils and Figure 4.14 for stiff or hard, fine grained soils.

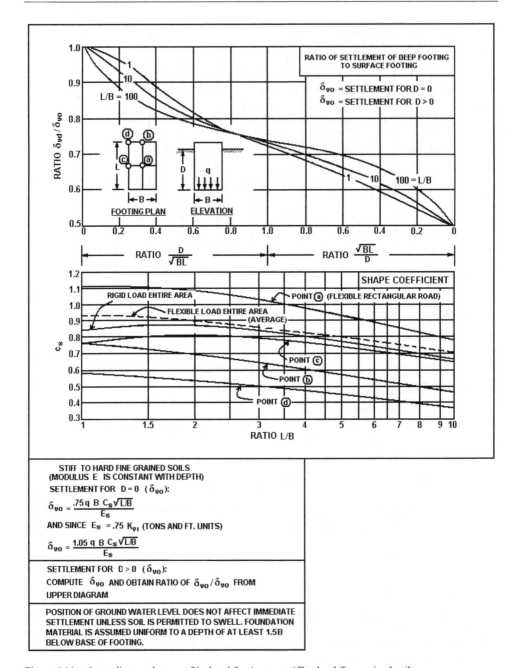

Figure 4.14.   Immediate settlement of isolated footings on stiff to hard fine grained soils.

REFERENCES

DeBeer, E.E. 1970. Experimental determination of the shape factors and the bearing capacity factors of sands. *Geotechnique*, 20 (4), London.

Hansen, J. 1970. Revised and extended formula for bearing capacity. *Danish Geotechnical Institutes Bulletin*, No. 28, Copenhagen.

Liao, S.S.C and Whitman, R.V. 1986. Overburden correction factor for SPT in sand. *Journal of Geotechnical Engineering, ASCE*, 3: 112.

Meyerhof, G.G. 1951. The ultimate bearing capacity of foundations. *Geotechniques* 2(4), London.

Prandtl, L. 1920. Uber die harte, plastischer Korper (On the hardness of plastic bodies). Nachr.Kgl.Ges.Wiss. Gottingen, Math. Phys. Kl.

Reissner, H. 1924. Zum erddruck-problem (The earth pressure problem). *Proceedings 1st International Congress on Applied Mechanics.*

Sanglerat, G. 1972. *The penetrometer and soil exploration*, London-New York-Elsevier.

Skempton, A.W. 1951. The bearing capacity of clays. *Proceedings Building Research Congress.*

Terzaghi, K. 1943. *Theoretical soil mechanics*. John Wiley. New York.

Terzaghi,K. and Peck, R.B. 1968. *Soil mechanics in engineering practice*, 2nd ed. New York: Wiley.

Vesic, A.S. 1975. Bearing capacity of shallow foundations. In H.F. Winterkorn and H.Y. Fang, eds., *Foundation Engineering Handbook*. New York: Van Nostrand-Reinhold.

CHAPTER 5

Driven Pile Foundation, Pile Driving Systems and Formulas

Bujang B.K. Huat
Department of Civil Engineering, University Putra Malaysia, Malaysia

Mageswaran Pavadai
R & A Geotechnics Sdn. Bhd., Kuala Lumpur, Malaysia

5.1 INTRODUCTION

The driven pile is the oldest and most widely used deep foundation system. Piles are normally columns made of wood, concrete, steel or plastic. They could also be composites such as concrete-filled steel pipe or plastic-steel composite, which are driven into the ground.

Wood or timber piles have been used for thousand of years, and continue to be a good choice for many applications. They are made from trunks of straight trees (such as the *bakau* or mangrove tree pile), and driven upside down with the larger diameter as the head, or it is sawn from a bigger log according to requirements and treated. Timber piles are usually designed to carry axial loads of 100 to 200 kN, while *bakau* piles are normally designed with a working axial load of 10 kN. Concrete piles are slender columns of precast (in some cases also prestressed) concrete driven into the ground. The piles usually come with a square or round cross-section, but other shapes such as triangular and octagonal are also available. They are typically 250 to 600 mm in diameter, 12 to 30 m long, with a working axial load of 450 to 3500 kN. Steel pile is another common type of pile, especially on projects that require high capacity foundations, and in marine applications due to their high tensile load carrying capacity (e.g. plastic composite steel pile). Because of their high strength and ductility, steel piles can be driven through hard soils and carry a large load. Steel piles have another advantage in that they are easy to splice. The common forms are steel piles of special rolled steel section, known as HP sections, or simple H piles, and steel pipes. The H piles are typically 15 to 50 m long and carry working axial loads of 350 to 1800 kN. Steel pipe piles are typically 200 to 1000 mm in diameter, 30 to 50 m long, and carry working axial load of 450 to 7000 kN. These piles are generally more expensive compared to the concrete piles.

In the case of tropical Malaysia, the development of pile systems to suit the varied tropical ground conditions has come a long way (Ting 1998). The earliest commonly used piles to support up to medium-rise structures in poor ground was the 300×300 mm and 375×375 mm square section reinforced concrete (RC) pile with grade 20 concrete, cast in situ as there was then no pile manufacturing facilities. For smaller structures and shallow depths (up to about 6 m) of poor ground with high ground water table, *bakau* (mangrove tree) which is a type of timber pile were used. Treated timber piles were later introduced. With the availability of modern pile manufacturing facilities, precast RC piles are now easily available. The production of driven precast-prestressed spun concrete pile, originally developed in Japan, is a fully developed industry in Malaysia, because of the ease in handling and its suitability in a wide range of formations found in Malaysia, i.e. from soft clay/loose sands to the harder formations. Steel H piles are also used. Figure 5.1 shows some of the commonly used driven piles in Malaysia.

Table 5.1 summarizes the structural strengths of precast concrete piles, adopted from the Malaysian Standard (MS1314 1993). Table 5.2 lists the basic structural requirements for precast concrete piles.

By virtue of transmission of load, driven piles can be classified as friction or end bearing piles. In friction pile, load is transmitted by the friction at the interface between the soil surface and its

(a) *Bakau* piles

(b) Precast concrete piles

(c) Spun piles

Figure 5.1.   Some commonly used driven piles in Malaysia.

surrounding soil as shown in Figure 5.2 (a). The pile in this case is not seated on a rock surface or hard soil layer. In the case of an end bearing pile, a large component of the load is transferred to the rock surface at the pile tip, (Figure 5.2 (b)).

A commonly used system for classifying piles is provided by the British Standards, BS 8004 (1986). In this system, a pile is classified according to its effect on the surrounding soils. Three main classes of piles are:

a. Large displacement pile
b. Small displacement pile
c. No displacement pile

Table 5.1.   Structural strengths of precast concrete piles (MS1314 1993).

Pile type	Pile size (mm × mm)	Crack moment (kN.m)	Ultimate moment (kN/m)	Max load (kN)	Remarks
R.C Piles Class 1	200 × 200	5.7	15	350	• Grade 40 concrete
	250 × 250	12.6	27	600	• 1.2% (min main reinf
	300 × 300	22.0	56	800	• Crack moment Mc at 0.20 mm
	350 × 350	35.0	76	1150	crack width
	400 × 400	50.0	111	1450	• Ultimate moment at zero axial load
Spun Piles Class A	250 mm	17	3	450	• Grade 55 concrete
	300	25	47	600	• Min effective stress 5 N/mm^2
	350	35	66	850	• Average loss in prestress 14%
	400	55	101	1100	• Crack moment Mc at 0.05 mm
	450	80	114	1300	crack width
	500	110	158	1600	
	600	180	265	2100	
	700	270	442	2800	
	800	400	606	3500	
	900	600	870	4300	
	1000	800	1150	5200	

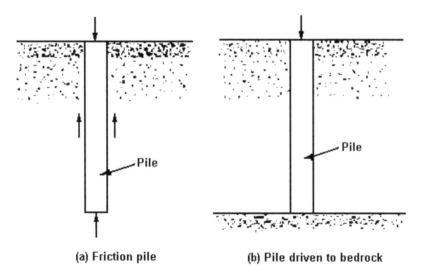

(a) Friction pile   (b) Pile driven to bedrock

Figure 5.2.   Driven pile foundation.

Driven piles are generally classified as large displacement piles because of their effect on the surrounding soils, except for H pile that can be classified as small displacement pile.

5.2   DESIGN OF DRIVEN PILE

The load carrying capacity of a driven pile (and so is the case for the drilled shaft or bored pile) can be determined by three methods, namely,

1. Analytical methods, which are based on the soil properties obtained from laboratory or in situ tests.

Table 5.2.   Basic structural requirements for precast concrete pile (MS 1314 1993).

Type & class of pile requirements	Precast reinforced concrete pile		Precast pretensioned concrete pile		Precast pretensioned spun concrete pile	
	Class 1	Class 2	Class X	Class Y	Class A	Class B
1. Grade of concrete	40	25	50	55	50	55
2. Minimum cement content (Kg/m)	400	320	450	450	450	450
3. Min longitudinal reinforcement	Min 1.2% of cross section of at least 4 bars	Min 1.0% of cross section of at least 4 bars	Min effective stress of N/mm²	Min effective stress of 7 N/mm²	Min effective stress of 5 N/mm²	Min effective stress of 7 N/mm²
4. Lateral reinforcement – at pile/toe	1.0% of gross volume over distance of 3 X pile width from each end.	0.6% of gross volume over distance of 3 X pile width from each end.	Min bar size of 6 mm diameter at 50 mm spacing over distance of 3 X pile width from each end.		Spiral hard drawn steel wire shall be not less than 4 mm diameter at a pitch of 50 mm over distance of 3 X pile size from each end.	
– at pile body	0.4% of gross volume spaced at not more than 1/2 X pile width	0.2% of gross volume space at not more than 1/2 X pile width	Spacing gradually increased from pile end, spacing not exceeding pile size less 60 mm		Spacing gradually increased from pile end, spacing not exceeding 100 mm.	
	• Min bar/wire size is 6 mm diameter • Min spacing 40 mm	• Min bar/wire size is 6 mm diameter • Min spacing 40 mm			• For pile size exceeding 600 mm, diameter of spiral hard drawn steel wire shall not be less than 6 mm	
5. Nominal cover	40 mm	40 mm	30 mm	30 mm	20 mm	20 mm
6. Minimum thickness of mild steel plate for joint (mm)	15 mm for pile size large than 250 square 12 mm for pile size 250 mm & smaller	15 mm for pile size large than 250 square 12 mm for pile size 250 mm & smaller	15 mm for pile size large than 250 square 12 mm for pile size 250 mm & smaller		15 mm for pile size larger than 250 diameter, 12 mm for pile size 250 mm	
7. Centering bar, (minimum anchorage length 200 mm and protruded length of 100 mm)	32 mm diameter centering bar for pile size larger than 250 mm square 25 mm diameter centering bar for pile size 250 mm & smaller		32 mm diameter centering bar for pile size larger than 250 mm square 25 mm diameter centering bar for pile size 250 mm & smaller			
8. Standard size	Square section of 200 mm, 300 mm, 350 mm & 400 mm		Square section of 200 mm, 250 mm, 300 mm, 350 mm & 400 mm		Cylindrical hollow section of external diameter of 250 mm, 300 mm, 350 mm, 400 mm, 500 mm, 600 mm, 700 mm, 800 mm, 900 mm, 1000 mm & 1200 mm	

Note 1. Spacing of lateral reinforcement shall be gradually changed from pile head/toe pile body.

2. Dynamic methods, which are based on the dynamic properties of pile driving or wave propagation, and,
3. Full-scale static load test on prototype foundations.

In this chapter we will describe the methods of estimating the load carrying capacity of piles using the analytical and dynamic methods, based on the dynamics of pile driving. The full-scale static load test and dynamic methods based on wave propagation will be described in another chapter of the book.

For the case of the analytical method, the load carrying capacity of a driven pile is designed either as a single pile or as a pile group.

Capacity of a single pile

Figure 5.3 shows a single pile subject to load, P and self-weight, W. The load is resisted by force at base, Q_b and shear force between the pile shaft and surrounding soil, R_s. Load P can be increased until the pile fail, at which instance the value of P becomes the ultimate load, P_f. Value of P_f depends on the size and length of the pile, and soil strength used in the calculation, R_{sf} and Q_{bf}. R_{sf} is the ultimate shear force (or skin friction) along the pile shaft, while Q_{sf} is the ultimate bearing force at the pile base.

Ultimate capacity of pile can be calculated from the following equation:

$$P_f + W = R_{sf} + Q_{bf} \tag{5.1}$$

Pile capacity in clay

The method usually used is the total stress analysis using undrained shear strength parameter. However, analysis using effective stress parameters can also be used.

Total stress analysis (α method)

Based on Figure 5.3, R_{sf} can be obtained from the following equation:

$$R_{sf} = A_s \times c_a \tag{5.2}$$

where $A_s = $ cross sectional area of pile shaft; $c_a = $ average adhesion.

Average adhesion

$$c_a = \alpha \times c_u \tag{5.3}$$

where α is an adhesion factor of value varying from 1.0 to 0.2, depending on soil undrained shear strength (higher value for low strength soil), type and length of piles, soil stratum, and the way the

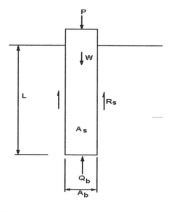

Figure 5.3.   Capacity of a single pile.

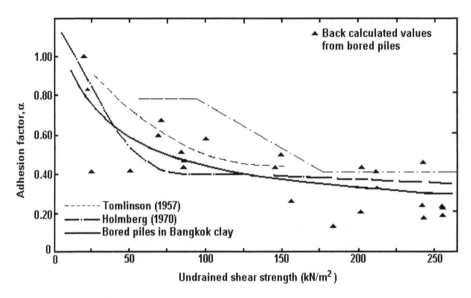

Figure 5.4.   Adhesion factor for clays (from Sambhandaraksa & Pitupakhorn 1985).

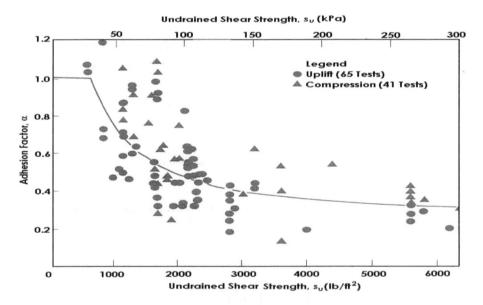

Figure 5.5.   Adhesion factor, α in drilled shafts (from Kulhawy & Jackson 1989).

pile is inserted into the soil, either by driving or drilling. Figure 5.4 shows the adhesion factors (α) for clay compiled by Sambhandaraksa & Pitupakhorn (1985) from their work in Thailand. The factors are found to be reliable for estimating the capacity of small driven piles, but for larger piles of 600 to 800 mm diameter, the capacity estimated needs to be enlarged by a factor of 1.2 (Balasubramaniam et al. 2004). Figure 5.5 shows the adhesion factor back calculated from instrumented load test of drilled shafts in clay. Value of c_u is average undrained strength of soil along the pile shaft.

From Equation (5.2) and (5.3), it is found;

$$R_{sf} = A_s \, \alpha \, c_u \tag{5.4}$$

Ultimate force at base, Q_{bf} is calculated using the following equation

$$Q_{bf} = A_b \times q_f \qquad (5.5)$$

where A_b = cross sectional area of pile base; q_f = ultimate bearing capacity at suitable depth.
   From the bearing capacity equation

$$q_f = N_c \times c_{ub} + \sigma_{vo}$$
$$= N_c \times c_{ub} + \gamma \times L \qquad (5.6)$$

where N_c = bearing capacity coefficient; σ_{vo} = overburden pressure; γ = soil unit weight; L = length of pile; c_{ub} = average undrained strength at pile base.
Hence

$$Q_{bf} = A_b(N_c \times c_{ub} + \gamma \times L) \qquad (5.7)$$

The ultimate capacity of the entire pile can then be calculated as follows:

$$P_f + W = R_{sf} + Q_{bf}$$
$$P_f + W = \alpha \times A_s \times c_u + A_b \times N_c \times c_{ub} + A_b \times \gamma \times L \qquad (5.8)$$

where $W = A_b \times L \times \gamma_c$ (γ_c is the unit weight of concrete) and if it is assumed that $\gamma_c \cong \gamma$, then:

$$W \cong A_b \times L \times \gamma \text{ and the above equation becomes}$$
$$P_f = \alpha \times A_s \times c_u + A_b \times N_c \times c_{ub} \qquad (5.9)$$

   It must be noted here that the value of P_f is the maximum load that the pile can carry. If $P > P_f$, the pile will fail, meaning settlement of the pile will exceed the allowable limit or stress at the pile base exceeds the soil bearing capacity.
   For soil with undrained strength, which increases with depth, the c_u along the pile shaft has to be distinguished from the c_u at the pile base. For pile shaft, c_u, is taken as the average c_u value along the shaft, whereas for the pile base, c_u value is taken as the average c_u over depth of $L + (2/3)B$, where B is the width of the pile base.
   Value of coefficient of bearing capacity, N_c can be obtained from the bearing capacity theory. For piles, if the ratio of its depth, L, with width, B, is far larger than 4 (i.e. $L/B \gg 4$), then the value of N_c normally used is 9.

Pile capacity in granular soil
For piles in granular soil, a large portion of the load is supported by the pile tip (base). However, the contribution of the pile shaft must also be taken into account. As in the case of piles in clay, pile capacity is calculated using effective stress analysis
   From Equation 5.1

$$P_f + W = R_{sf} + Q_{bf}$$

For granular soil

$$R_{sf} = A_s \times k \times \sigma'_{vo}. \tan \delta' \qquad (5.10)$$

where A_s = area of pile shaft; σ'_{vo} = average effective overburden pressure along the pile; k = earth pressure coefficient; and δ' = angle of friction between soil and pile.
   Value of k actually varies with depth. At the pile head, its value is about the same as Rankine coefficient of passive (k_p), while at the pile tip, its value is similar to k_o, coefficient of earth pressure at rest. This coefficient also depends on method of pile construction. Table 5.3 shows the values of k for a number of pile types.
   Table 5.4 shows the value of k for the various pile construction methods.

Table 5.3.   Value of k for various types of piles (from Broms 1965).

Pile type	Value of k	
	Loose soil	Dense soil
Steel	0.5	1.0
Concrete	1.0	2.0
Wood	1.5	3.0

Table 5.4.   (a) Value of k for various classes of piles (from Das 1984).

Method	Value of k	
	Upper limit	Lower limit
Drilled	k_o	–
Driven-small displacement	k_o	1.4 k_o
Driven-large displacement	–	1.8 k_o

Note: $(k_o = 1 - \sin \phi')$ (where ϕ' is soil effective angle of friction).

(b) Value of k (after Kulhawy et al. 1983).

Foundation type and method of construction	k/k_o
Pile-jetted	0.5–0.7
Pile-small displacement, driven	0.7–1.2
Pile-large displacement, driven	1.0–2.0
Drilled shaft-built using dry method with minimal sidewall disturbance and prompt concreting	0.9–1.0
Drilled shaft-slurry construction with good workmanship	0.9–1.0
Drilled shaft-slurry construction with poor workmanship	0.6–0.7
Drilled shaft-casing method below water table	0.7–0.9

Value of δ' is usually smaller than ϕ' as shown in Table 5.5. These values are considered as safe values. The ϕ' is the soil (soil-to-soil) angle of friction.

The value of δ' can be estimated based on field tests such as the Standard Penetration Test (SPT). Figure 5.6 shows an example of a relationship between uncorrected SPT N value and δ' for clayey sand complied by Sambhandaraksa & Pitupakhorn (1985) from their work in Thailand.

Figure 5.7 shows a plot of average shaft friction $(\tau_s = R_{sf}/A_s)$ with average effective overburden pressure (σ'_{vo}) from pile load test in clayey sand compiled by Balasubramaniam et al. (2004). The average ratio of shaft friction to effective overburden pressure (β) is about 0.35.

Table 5.6 lists some empirical values of angle of friction, ϕ, relative density, DR, and bulk density of granular soils based on the SPT, N and Dutch cone (deep sounding) resistance values.

The following simplified relations may also used to estimate the skin friction along the pile shaft:

$$R_{sf} = 2N \tag{5.11}$$

$$R_{sf} = 0.05 C_{kd} \tag{5.12}$$

The R_{sf} is not greater than 100 kPa for driven or drilled shaft in residual or cohesive soil, N is the SPT value, and C_{kd} is the cone resistance from static Dutch cone test.

Table 5.5.   (a) Value of δ' for pile (from Aas 1966).

Pile type	δ'
Steel	20°
Concrete	$3/4\phi'$
Wood	$2/3\phi'$

(b) Value of δ' (after Kulhawy et al. 1983).

Foundation type	δ'
Rough concrete	$1.0\phi'$
Smooth concrete (i.e. precast pile)	0.8–$1.0\ \phi'$
Rough steel (i.e. step-taper pile)	0.7–$0.9\phi'$
Smooth steel (i.e. pipe pile or H-pile)	0.5–$0.7\ \phi'$
Wood (i.e. timber pile)	0.8–$0.9\ \phi'$
Drilled shaft built using dry method or with temporary casing and good construction techniques	$1.0\ \phi'$
Drilled shaft built with slurry method (higher values correspond to more careful construction methods)	0.8–$1.0\ \phi'$

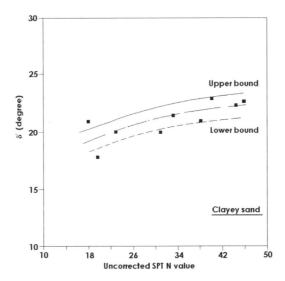

Figure 5.6.   Relationship between uncorrected SPT N value and δ' of clayey sand (Sambhandaraksa & Pitupakorn 1985).

The capacity of pile tip/base can be obtained from the following equation:

$$Q_{bf} = A_b \times q_f \tag{5.13}$$

For sand and granular soil

$$q_f = \sigma'_{vo}(N_q - 1) \tag{5.14}$$

With that, the ultimate capacity of the entire pile is

$$P_f = A_s \times k \times \sigma'_{vo} \times \tan \delta' + A_b \sigma'_{vo}(N_q - 1) \tag{5.15}$$

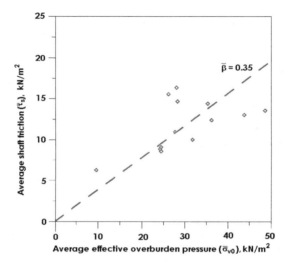

Figure 5.7.   Estimation of β from pile load test.

Table 5.6.   Empirical values ϕ, DR, and bulk density of granular soils base on the SPT, N and Dutch cone (deep sounding) resistance values.

Description	V. Loose	Loose	Medium	Dense	V. Dense
SPT (N)	<4	4–10	10–30	30–50	>50
Relative Density, DR	<0.2	0.2–0.4	0.0–0.6	0.6–0.8	>0.8
Bulk density (kN/m^3)	11–16	14–18	17–20	17–22	20–23
DS Cone Re (N/cm^2)	<20	20–40	40–120	120–200	>200
ϕ'	<30°	30°–35°	35°–40°	40°–45°	>45°

NOTES:
1. ϕ'– lower limit for uniform clean sands and upper limit range for well-graded sands. For silty sand, ϕ should be reduced by 5°.
2. SPT (N') corrected for depth and for fine saturated sands.
3. $N' = 15 + \frac{1}{2}(N - 15)$ if $N > 15$ and the sand is very fine or silty and saturated.
4. DS Cone Re = Deep Sounding Cone Resistance Value.

Note that in Equation 5.15, unit weight is assumed to be equal to unit weight of pile. Because N_q is dependent on ϕ', a sample has to be obtained and tested for ϕ'. For sand and granular soils, it is difficult to obtain their undisturbed samples. Because of this, an in situ test is usually performed to obtain ϕ'. Two in situ test methods normally used are the standard penetration test (SPT) and cone penetration test (CPT). Some of these correlations have been presented earlier in this chapter. Figure 5.8 shows a plot of the bearing capacity factor, N_q.

It is of interest to note that the following simplified relation may also used to estimate q_f

$$q_f = k_b N \tag{5.16}$$

where k_b is an empirical factor, and equals to 100, 250 and 400 for clay, silt and sand respectively, and N is the SPT value. A value of k_b equal to 100 (kPa) is also suggested for drilled shaft.

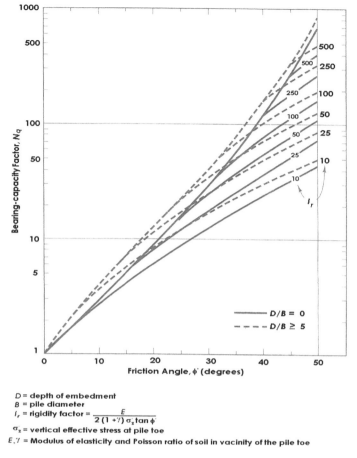

D = depth of embedment
B = pile diameter
I_r = rigidity factor = $\dfrac{E}{2(1+\gamma)\sigma_z \tan\phi'}$
σ_z = vertical effective stress at pile toe
E,γ = Modulus of elasticity and Poisson ratio of soil in vacinity of the pile toe

Figure 5.8.   Bearing capacity factor, N_q. (from Kulhawy et al. 1983).

Pile design load

Ultimate capacity of pile calculated using Equations 5.9 and 5.15 is the maximum load that a pile can carry before it fails. Actually it is difficult to really determine the ultimate capacity of a pile since there is no clear reference point when the pile has failed. In any case, there are two definitions that are normally given to ultimate pile capacity. They are:

a. Load when soil resistance has been fully mobilized.
b. Load when the pile is displaced (downward) by distance 10% the diameter dimension (smallest measurement) of the pile end.

To obtain design load or P_w, P_f is divided by factor of safety, F, that is

$$P_w = P_f/F \qquad (5.17)$$

If P_f is obtained from load test (say constant rate penetration test), value of $F = 2$ is usually used.

If ultimate capacity, P_f, is based on soil parameters obtained from laboratory test, $F = 3$ is used.

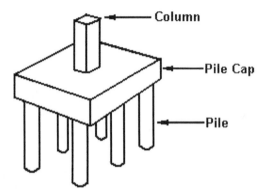

Figure 5.9.   Pile group with a typical pile cap.

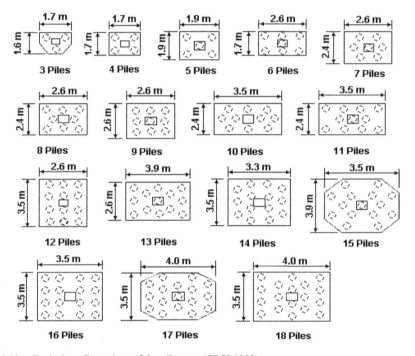

Figure 5.10.   Typical configurations of the pile caps (CRSI 1992).

Alternatively the following equations are used:

$$P_w = R_{sf}/2 + Q_{bf}/3 \tag{5.18}$$

or

$$P_w = (R_{sf} + Q_{bf})/2.5 \tag{5.19}$$

Capacity of pile group

To support larger loads, pile group is normally used. A pile cap, which is normally a reinforced concrete member that is similar to a pad footing, is constructed on top of the group as shown in Figure 5.9. In addition the effect of eccentricities is much less significant as the column is supported on three or more piles. Typical configurations of the pile caps are shown in Figure 5.10. Sometimes

Table 5.7. Minimum distance for pile group (from BS 8004: 1986).

Pile type	Minimum distance between pile
Friction	Pile circumference
End bearing	Twice smallest width of the pile

the individual pile caps are connected with grade beams, which are structural beams embedded in the ground.

Distance between piles
Distance between piles has to be accounted for as it determines whether the pile is acting alone or in a group. If the distance between the piles is large, then the pile is acting alone. If the distance between the piles is too far apart, the cost of constructing the pile cap might be too high. On the other hand, if the pile is too close, problem of heaving at the soil surface may arise. Hence a suitable distance needs to be chosen so that the heave problem can be minimized. The minimum distance is dependent on the method of pile construction (either driven or drilled) and the bearing capacity of the soil.

As reference, BS 8004: 1986 determines minimum distance for a number of piles as shown in Table 5.7.

Calculation for ultimate capacity of group pile is not easy, because of the complexity in interaction between the piles and their surrounding soil. In the pile group, stress overlaps between the pile and its surroundings. Due to this, the ultimate capacity of the pile group is reduced. The optimum condition is when the pile is spaced in such as way that the capacity of the pile group is the summation of each of the individual piles.

Group pile efficiency, η, can be defined as

$$\eta = \frac{P_{fg}}{\Sigma P_f} \tag{5.20}$$

where P_{fg} = ultimate capacity of pile group; P_f = ultimate capacity of single pile.

Group pile in clay
Figure 5.11 shows a group pile and the interaction between the pile and the ground. If the pile is assumed to fail as a group, then the soil around and at the base of the pile will fail. This is called a block failure.

Skin friction of the pile group is

$$R_{sg} = 2L(B + W)c_u \tag{5.21}$$

Bearing force at the base of the pile group is

$$Q_{bg} = N_c \times c_{ub} \times B \times W \tag{5.22}$$

Hence, ultimate capacity of the pile group is

$$P_{fg} = 2L \times c_u(B + W) + N_c \times c_{ub} \times B \times W \tag{5.23}$$

For firm clay, the lowest of the ultimate capacity from the two following calculations are chosen, that is

a. Capacity of a single pile multiplied by number of piles in the group, or,
b. Capacity calculated based on assumption of block failure.

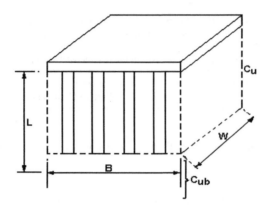

Figure 5.11.   Pile group in clay.

Pile group in sand
Calculation for bearing capacity of a pile group in granular soil is similar to calculation for pile group in clay.
   Skin friction forces

$$R_{sg} = A_{sg} \times k \times \sigma'_{vo} \times \tan \delta' \tag{5.24}$$

where, A_{sg} = circumference area of pile group; k = earth pressure coefficient; δ' = angle of friction between pile and soil; and σ'_{vo} = average effective overburden pressure.
   Bearing force at base of group is

$$Q_{bg} = A_{bg} \times q'_f \tag{5.25}$$

where, A_{bg} = circumference area of pile group; and q'_f = bearing capacity at base = $\sigma'_{vo} (N_q - 1)$.
   Ultimate bearing capacity is the summation of total skin resistance and bearing force at base, that is

$$P_{fg} = R_{sg} + Q_{bg} \tag{5.26}$$

Usually P_{fg} is larger than the total ultimate capacity of single pile inside the group. This means that efficiency of greater than one may be obtained. This condition exists when the granular soil become denser when the piles are driven into it. The compaction increases the bearing capacity of sand in between the pile shafts.
   But in design, the ultimate capacity of the pile group in sand is considered as unity. If $\eta < 1$, then,

$$P_{fg} = \eta \Sigma P_f \tag{5.27}$$

And if $\eta > 1$, then

$$P_{fg} = \Sigma P_f \tag{5.28}$$

Negative skin friction

Negative skin friction around a pile shaft will act downward (in the same direction as load) when a pile is inserted into clay above a firm layer as shown in Figure 5.12. Negative skin friction will occur if

i. Pile is inserted into soil which is experiencing consolidation.
ii. Effective stress increases after the pile is inserted either due to filling or lowering of ground water table.

   Negative skin friction (R_s) can be calculated using the following equation:

$$R_s = \Delta A_s \beta \sigma'_v \tag{5.29}$$

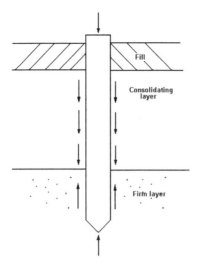

Figure 5.12. Negative skin friction.

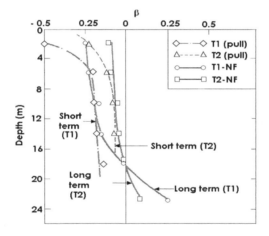

Figure 5.13. Variation of β-parameter with depths during negative skin friction (Indranatna et al. 1992).

where, ΔA_s = area of pile shaft in clay layer; σ_v' = average effective vertical stress at the said depth and β = coefficient.

Figure 5.13 shows an example of the variation of β with depths during negative skin friction based on a full scale tension pile test. The maximum negative skin friction developed was about $0.25\,\sigma_v'$ ($\beta = 0.25$), with an average value of about $0.2\,\sigma_v'$.

5.3 PILE DRIVING SYSTEMS

Pile driving hammers

Conventional pile driving techniques includes dropping of drop or diesel hammer (also known as impact hammer) to drive in the pile. Pile driving therefore has the potential of being the greatest source of ground vibration and noise generated by piling activities during a construction, in addition to air pollution in the case of a diesel hammer. Current design code places limits on ground vibration

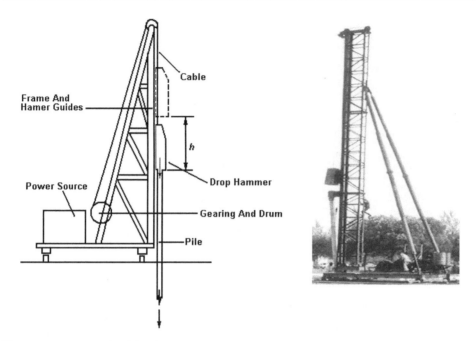

Figure 5.14.   Power assisted drop hammer.

and noise generated by these piling activities. These limits are intended to prevent disturbances to humans and damage (both cosmetics and structural) to nearby buildings, particularly in the urban environment. For this, a quieter pile driver, using the hydraulic and vibratory system, is available.

Conventional impact hammers include drop, single-acting and double acting steam, compressed air or diesel hammer. Figure 5.14 shows an example of a power assisted drop hammer. The hammer weight is hung from a cable running over a pulley. The whole arrangement is supported on a strong frame or leader. The hammer is released by a manual trip and drops under free fall. This method is fairly slow, but is simple, requires little maintenance and specialist contractors and is commonly used for all types of pile driving. Good practice requires that the weight of the hammer be equal to the weight of the pile. However, care must be taken when heavy hammers are used to drive light piles into a hard stratum as over-driving leads to overstressing of the pile. For heavy hammers, the drop can be reduced or if the depth of the hard stratum is variable, it may be desirable to use a lighter hammer. The height of the pile driving frame is to be sufficient to conveniently handle the main length of the pile to be driven. The pile is to be properly constrained during driving so that deviation from true alignment is not excessive.

Single acting steam or compressed air hammers drive piles in a similar manner to the drop hammer, but the hammer is raised by steam or air rather than by winching. An example of a typical single acting steam hammer is shown in Figure 5.15. A higher striking rate is achieved with this hammer, thus delivering heavy blows (high impact energy) to the pile. A double acting steam or compressed air hammer employs steam or air that is admitted to the upper and lower cylinders alternately by means of valves actuated by piston. In this way, both the free fall impact and additional energy from the release of compressed air into the upper cylinder are obtained on the downward stroke. By switching the air supply to the lower cylinder, the piston is raised and the air in the upper cylinder is expelled, ready to repeat the cycle. Double acting hammer runs at a considerably greater speed than the single acting hammer, which is particularly advantageous in overcoming static friction in granular soils.

The diesel hammer (Figure 5.16) has proved to be a successful, and in many cases a better alternative to the drop and single acting steam or compressed air hammer, because the available

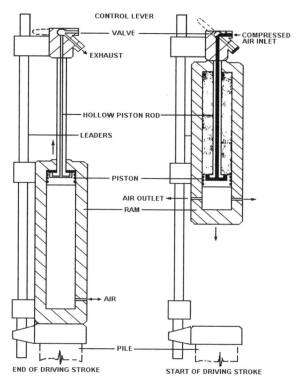

Figure 5.15.   Single acting air or steam hammer (after Harris 1994).

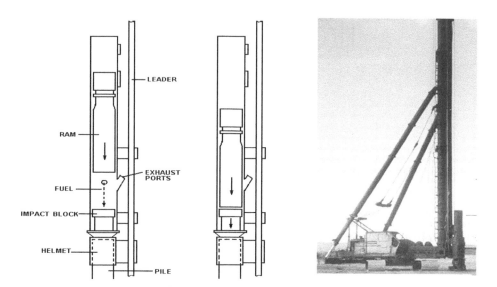

Figure 5.16.   Diesel hammer.

energy per blow is about double for a comparable weight of ram. The hammer is similar to a diesel internal combustion engine. The diesel hammer is most suitable where a very heavy blow (high impact energy) is required. The hammer develops its maximum energy under hard driving conditions, and may be difficult to operate under soft conditions, because of lack of proper combustion

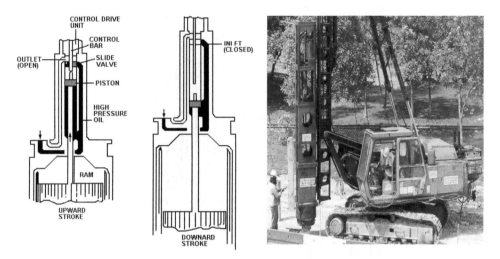

Figure 5.17.   Hydraulic impact hammer (diagram after Harris 1994).

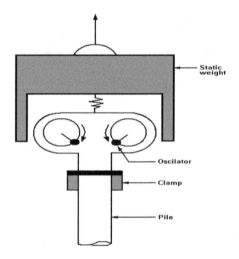

Figure 5.18.   Vibratory hammer.

or insufficient hammer rebound. This hammer has been popularly used for many years, but as its exhaust is a source of air pollution, air quality regulations restrict its use in some areas.

The hydraulic impact hammer (Figure 5.17) is gaining popularity with the increase availability of hydraulic powering of modern construction equipment. The working principle is similar to the double acting air or steam hammer, but hydraulic fluid is used in this case both to raise the ram and provide additional driving force to gravity.

A vibratory hammer (Figure 5.18) is not a hammer in the same sense as those described earlier. It uses rotating weights to create vertical vibrations. When combined with a static weight, these vibrations force the pile into the ground. These hammers are most effective when used with piles driven into sandy soils. They operate more quickly and with less vibration and noise than conventional impact hammers. However, they are ineffective in clays or soils containing obstructions such as boulders.

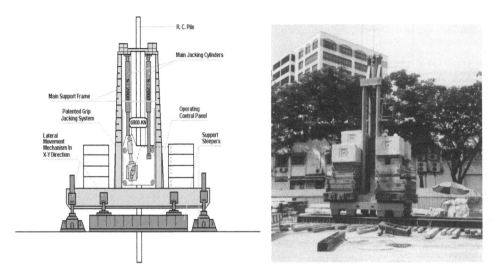

Figure 5.19.   Jack in piling system (diagram courtesy of G-Pile Systems Sdn. Bhd. Malaysia).

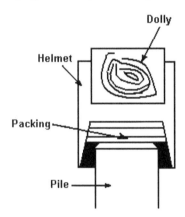

Figure 5.20.   Pile helmet, dolly and packing.

The hydraulic jack in piling system (Figure 5.19) is also available nowadays. The system utilizes a patented hydraulic grip jacking technique to inject the pile into the ground. Low noise, vibration and pollution are the main advantage of this system, particularly for urban and residential sites.

Pile helmets, splices and shoes

A pile helmet (Figure 5.20) is required to distribute the blow from the hammer evenly to the head of the pile, to cushion the blow and to protect the pile itself. A dolly is placed in a recess in the helmet and acts to cushion the blow. A packing plate is also placed between the pile head and the underside of the helmet, to act as cushion, especially in the case of concrete piles that are likely to suffer damage from the force of the hammer impact.

Driven piles when necessary are spliced. Figure 5.21 shows typical splices for timber, steel and concrete piles.

Figure 5.22 shows precast concrete piles fitted with the conventional steel pointed shoes. A flat-ended pile is equally suitable as pointed ended piles in many situations. In the case of piles seated on rock or hard stratum, an Oslo shoe (Figure 5.23) would be suitable.

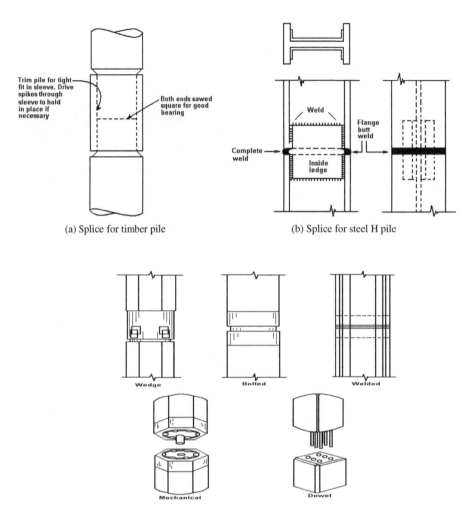

(a) Splice for timber pile

Trim pile for tight fit in sleeve. Drive spikes through sleeve to hold in place if necessary

Both ends sawed square for good bearing

(b) Splice for steel H pile

Weld

Flange butt weld

Complete weld

Inside ledge

Wedge

Bolted

Welded

Mechanical

Dowel

(c) Splices for concrete piles (after Precast/Prestressed Concrete Institute)

(d) Welding of a precast concrete pile

Figure 5.21. Splices for piles.

Figure 5.22.   Pointed pile shoes.

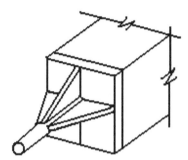

Figure 5.23.   Oslo shoe.

5.4   PILE DRIVING FORMULAS

When driving piles, the blow count, which is the number of hammer blows required to drive the pile for a specified distance, is usually monitored. The blow count is expressed in units of blow/ft or blow/250 mm. The most important value is the blow count for the last foot (or 250 mm). Figure 5.24 shows an example of a form used for the pile driving record.

Intuitively, it can be expected that piles difficult to drive have more load carrying capacity that those that are easier to drive. Many empirical correlations have been therefore developed to relate the hammer weight, blow count and other factors with the static load capacity of driven piles. These relationships are collectively known as pile driving formulas.

The basic relationship common to all pile driving formulas is

$$P_a = \frac{W_r h}{sF} \tag{5.30}$$

where P_a = allowable axial load capacity; W_r = weight of ram; h = hammer stroke (the distance the hammer falls); s = pile set (penetration) per blow at the end of driving and F = factor of safety.

One of the most popular pile driving formulas is the *Engineering News Formula*:

$$P_a = \frac{W_r h}{F(s + c)} \tag{5.31}$$

Coefficient c is taken as $c = 1$ in (25.4 mm) for drop hammer, and $c = 0.1$ in (2.54 mm) for single acting hammers with factor of safety, $F = 6$.

PILE – DRIVING RECORD

Project:	Location of pile test Drg No: ....
.... Contract No: .... Contractor:	Driving Sequence No: .... Date driven: .... Entries made by: .... Checked by: ....

	Type of pile:	Size:		Length:		Weight:	
Date of casting		Details of extension					
Type of pile frame : Type of hammer :			Height of frame: Weight of hammer :				
R.L of ground: .... R.L of top of driven pile:			Final set:				

R.L of cut-off level: .... R.L of toe of driven pile: .... Length of pile from cut-off level:				Measured temporary compression: .... Type of driving head:				
Penet-ration (ft)	Height of hammer drop (ft)	No. of Blows	Remarks	Penetra-tion (ft)	Height of hammer drop (ft)	No. of blows		Remarks

Figure 5.24.   Typical form for pile driving record.

A more recent form of the *Engineering News Formula* called the *Modified Engineering New Formula* is written as follows:

$$P_a = \left(\frac{e_h W_r h}{s + c} \right) \left(\frac{W_r + n^2 W_p}{W_r + W_p} \right) \cdot \frac{1}{F} \qquad (5.32)$$

where e_h is the hammer efficiency, depending in hammer type as below. W_p is the weight of the pile including the pile cap. Coefficient c is taken as 1 in. (25.4 mm) for drop hammer, and $c = 0.1$ in (2.54 mm) for a single acting hammer.

Type	Efficiency, e_h
Drop hammer	0.75–1.00
Single-acting hammers	0.75–0.85
Double-acting hammer	0.85
Diesel hammers	0.85–1.00

n is the coefficient of restitution and given as follows depending on the pile helmet and cushion.

Representative values of coefficient of restitution for use in the dynamic pile-driving equation (after ASCE 1941).

Material	n
Broomed wood	0
Wood piles (non-deteriorated end)	0.25
Compact wood cushion on steel pile	0.32
Compact wood cushion over steel pile	0.40
Steel-on-steel anvil on either steel or concrete pile	0.50
Cast-iron hammer on concrete pile without cap	0.40

There are also several other pile driving formulas such as the Canadian National Building Code formula, the Danish formula, the Gates formula, the Janbu formula, the AASHTO formula, the Navy-McKay formula, and the Pacific Coast Uniform Building Code formula.

At sites where a full-scale load test is not available (such as for small projects), pile driving formulas may be used to access the static load carrying capacity of driven piles. Each pile has a required load capacity that corresponds to a certain minimum acceptable blow count. Therefore the pile is driven until it reaches the specific blow count (refusal). In large projects where static load test is usually carried out, the pile driving formulas may be modified to match the load test results. This custom formula is then applied to other piles at the site, and thus provides a mean of extrapolating the load test results, and for purposes of construction control.

REFERENCES

Aas. G. 1966. *Baereevne Au Peler 1 Frisksjonsjordarter*. Norwegian Geotechnical Institute. Oslo.

ASCE 1941. Pile driving formulas. *Proceedings ASCE*. 67(5): 853–866.

Balasubramaniam, A.S. 1991. Evaluation of pile foundation works for Don Muang tollway project. Internal report. Asian Institute of Technology.

Balasubramaniam, A.S., Phienwej, N., Gan, C.H. & Oh, N.Y. 2004. Piled foundations and basement excavations for tall building in Bangkok subsoils. *Proceedings Malaysian Geotechnical conference*. Malaysia: Kuala Lumpur. 89–107.

BS 8004, 1986. British Standard Code of Practice for Foundation, BSI, London.

Broms, B.B. 1965. Methods of calculating the ultimate bearing capacity of piles: a summary. *Sols-Soils*, 5.

CRSI 1992. CRSI handbook. Concrete Reinforcing Steel Institute. Schaumburg, IL.

Das. B.M. 1984. *Principles of foundation engineering*. Brooks/Cole Eng. Div. California.

Harris, F. 1994. Modern construction and ground engineering equipment and methods. Longman Scientific and Technical, UK.

Indranatna, B., Balasubramaniam, A.S., Phamvan, P. & Wong, Y.K. 1992. Development of negative skin friction on driven piles in soft Bangkok clay. *Canadian Geotechnical Journal*. 29(3). 393–404.

Kulhawy, F.H., Trautmann, C.H., Beech, J.F., O'Rourkee, T.D., McGuire, W., Wood, W.A. & Capano, C. 1983. Transmission line structure foundations for uplift compression loading. Report EL-2870. Electric Power Research Institute. Palo Alto, CA.

Kulhawy, F.H. & Jackson, C.S. 1989. Some observations of undrained side resistance of drilled shaft. Foundation Engineering: Current Principles and Practises. ASCE. 1011–1025.

MS 1314. 1993. Precast concrete pile. Part 1: General requirement and specifications. Malaysian Standards.

Sambhandaraksa, S. & Pitupakhorn, W. 1985. Prediction of presstressed concrete pile capacity in stiff clay and clayey sand. *Proceedings of 8th South East Asian Geotechnical Conference.* Malaysia: Kuala Lumpur. 358–363.

Ting, W.H. 1998. A survey of pile systems in Malaysia – past and present. Foundation Course. The Institution of Engineers Malaysia. Malaysia: Sarawak. October.

CHAPTER 6

Drilled Shafts Foundation

Bujang B.K. Huat
Department of Civil Engineering, University Putra Malaysia, Malaysia

Mageswaran Pavadai
R & A Geotechnics Sdn. Bhd., Kuala Lumpur, Malaysia

6.1   INTRODUCTION

A drilled shaft foundation is a deep foundation that is constructed by drilling a large hole in the soil and subsequently filling it with concrete. The purpose of a drilled shaft is to transfer large axial or lateral loads to the supporting soil. The fundamental difference between driven piles and drilled shafts is that driven piles are prefabricated members driven into the ground, whereas drilled shafts are cast in place. Large size drilled shafts can be executed and inspected with a degree of care that cannot be attained in the case of small diameter shafts. It is also of interest to note that engineers and contractors use the following terms interchangeably to describe this class of deep foundations, namely drilled pier, pier, bored pile, and cast in place pile. Sometimes a drilled shaft is also referred to as caisson, which is confusing and should ideally be avoided. The word caisson, means box when applied to a foundation. It describes a prefabricated hollow box or cylinder that is sunk into the ground to some desired depth, and then filled with concrete. Drilled shafts are not the same as certain other methods that involve cast in place piles such as the auger cast pile and pressure injected footing.

The advantages of this type of foundation over other similar types are as follows:

a. The construction process generates less noise and vibration, both of which are especially important when working near existing buildings.
b. Engineers can observe and classify the soils excavated during drilling and compare them with the anticipated soil conditions.
c. Designers can easily change the diameter or length of the shaft during construction to compensate for unanticipated soil conditions.
d. The foundation can penetrate through soils with cobbles or boulders, especially when the shaft diameter is large. It is also possible to penetrate many types of bedrocks.
e. It is usually possible to support each column with one large shaft instead of several piles, thus eliminating the need for a pile cap.

The disadvantages, however, are:

a. Successful construction is very much dependent on contractor's skills. Poor workmanship can produce a weak foundation that may not be able to support the design load.
b. Shaft construction removes soil from the ground. This causes reduction of lateral soil pressure thus reducing skin friction.
c. The shaft construction loosens the soil beneath the pile tip thus giving a lower end bearing.
d. Full-scale load tests are usually very expensive.

In Malaysia, a drilled shaft is commonly referred to as bored pile. Such a shaft is typically 0.6 to 1.6 m in diameter, and can be constructed to depths in excess of 60 m. Figure 6.1 shows an example of the structural reinforcement and structural load of the drilled shaft.

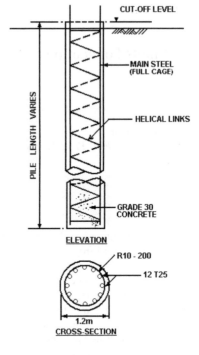

Pile diameter (mm)	Safe structural load*	
	(kN)	(ton)
800	3490	350
1000	5840	550
1100	6625	665
1200	7970	800
1420	10,960	1100
1500	12,450	1250
1600	14,015	1407

*After Chan 2004.

Figure 6.1.   Reinforcement details and safe structural loads of drilled shafts in compression.

6.2   DESIGN OF A DRILLED SHAFT

Axial load carrying capacity of drilled shafts

The primary load on a drilled shaft is usually axial, either downward or uplift. Therefore, it is necessary to select a diameter and length that produce sufficient skin friction and end bearing capacities. There are several methods of determining the axial capacity of drilled shaft. The important methods are:

a. Full-scale load test
b. High strain dynamic impact test
c. Presumptive bearing pressure
d. Analysis based on soil properties.

The most common method of designing a drilled shaft is based on soil properties obtained from field and laboratory tests. The analyses are similar to those of driven piles and are subject to similar uncertainties. End bearing and skin friction resistances are computed, though the methods of computing these factors are different because of their construction dissimilarities.

Cohesionless soil
End bearing: The unit end bearing capacity for drilled shafts in cohesionless soils will be less than that for driven piles because of the following:

a. The soil is disturbed by the auguring process.
b. The soil compression that occurs below driven piles is not present.
c. There is a temporary stress relief while the hole is open.
d. The diameter and depth of influence of the drilled shaft are greater.

As these and other factors are not well defined, simple and conservative empirical formula seem to represent an appropriate level of sophistication. O'Neil and Reese (1999) suggested the following formula:

$$q'_e = 57.5N_{60} \leq 2900\,\text{kPa} \tag{6.1}$$

where, q'_e = end bearing resistance (kPa); N_{60} = mean SPT N-value between toe and a depth of $2B_b$ below the toe; D = depth to the bottom of the shaft and B_b = diameter of the shaft base.

For shafts larger than 1200 mm in diameter, the value of q'_e is to be factored as follows:

$$q'_{er} = q'_e(1200\,\text{mm}/B_b) \tag{6.2}$$

Cohesive soils
End bearing: In the case of cohesive soil, O'Neil & Reese (1999) suggested the following equation for estimating the end bearing of a drilled shaft:

$$q'_e = N^*_c \cdot S_u \tag{6.3}$$

where q'_e = end bearing resistance (kPa); N^*_c = bearing capacity factor (6.5, 8.0, 9.0 at S_u equal to 25 kPa, 50 kPa and ≥ 100 kPa respectively); D = depth to the bottom of the shaft; S_u = undrained shear strength in the soil between the base of the shaft and a distance $2B_b$ below the base and B_b = diameter of the shaft base.

If the base diameter is more than 1900 mm, the value of q'_e could produce settlement greater than 25 mm (1 inch), which would be unacceptable for most buildings. To keep settlements within tolerable limits, O'Neil & Reese (1999) suggested the value of q'_e be reduced to q'_{er} as follows:

$$q'_{er} = F_r q'_e \tag{6.4}$$

$$F_r = 2.5/(\psi_1 B_b + 2.5\psi_2) \leq 1.0 \tag{6.5}$$

$$\psi_1 = 0.28B_b + 0.083(D/B_b) \qquad \text{(in SI units)} \tag{6.6}$$

$$\psi_2 = 0.065\sqrt{(S_u)} \qquad \text{(in SI units)} \tag{6.7}$$

Skin friction
Skin friction resistance in cohesive soils in the case of a drilled shaft may be evaluated using either α method or β method.

In the case of the α method, the ultimate skin friction is given by

$$f_s = \alpha S_u \tag{6.8}$$

where, α = adhesion factor; and S_u = undrained shear strength (cohesion) soil along the shaft.

Figure 6.2 shows the back-calculated values of α for drilled shafts obtained from instrumented load tests. Skin friction resistance in the upper 1.5 m of the shaft, and at the base of the shaft (at height equal to the diameter of the shaft from the base), is usually ignored.

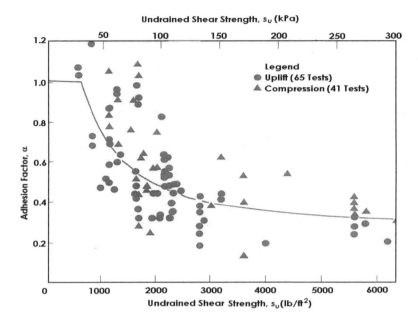

Figure 6.2.   Adhesion factor, α, in drilled shafts (from Kulhawy & Jackson 1989).

In the case of the β method, the following equation as suggested by Burland (1983) is generally used:

$$f_s = \Sigma \beta \sigma'_v \qquad (6.9)$$

where f_s is the skin friction along the pile shaft, σ'_v is the effective midpoint vertical stress of the soil layer, and β is a coefficient. For a drilled shaft in sand with $N_{60} \geq 15$, O'Neill and Reese (1999) recommend the following equation for β, and limit f_s to 190 kPa.

$$\beta = 1.5 - 0.245\sqrt{z} \quad 0.25 \leq \beta \leq 1.20 \qquad (6.10)$$

where z is the depth to midpoint of soil layer (in m). In the case of $N_{60} < 15$, the value of β obtained from Equation 6.10 is reduced by ratio of $N_{60}/15$.

Meyerhof (1976) suggested the following formula for estimating the ultimate load carrying capacity of a single drilled shaft in granular soil. This method is widely used in many countries, including Malaysia.

$$Q_u = N_s \cdot \frac{A_s}{k_s} + k_b \cdot N_b \cdot A_b \qquad (6.11)$$

where Q_u = ultimate axial load capacity of the shaft (tons); N_s = SPT value along the shaft; A_s = perimeter area of the shaft (ft^2); k_s = empirical design factor for skin friction; k_b = empirical design factor for end bearing; N_b = SPT value at the base; and A_b = cross-sectional area of the base.

The allowable load, Q_a, is

$$Q_a = \frac{Q_u}{2.5} \qquad (6.12)$$

or

$$Q_a = \frac{N_s \cdot A_s}{k_s F_s} + k_b \cdot N_b \cdot A_b \cdot \frac{1}{F_b} \qquad (6.13)$$

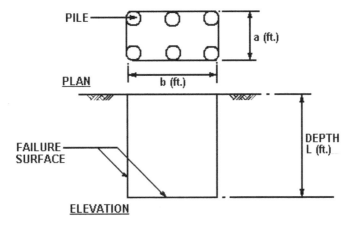

Figure 6.3.   Shaft group.

where F_s is the partial factor of safety for the shaft; and F_b is the partial factor of safety for the base

Chan (2004) suggested the following values,

F_s 1.5 for shaft of 0.8 to 1.6 m diameter and
F_b varied from 3.5 for a shaft of 0.8 m diameter to 6.3 for shaft of 1.6 m diameter

For a shaft group, the ultimate load carrying capacity of the group/block, Q_{ug}, is given as follows:

$$Q_{ug} = \sum \left[(2a + 2b) \cdot \Delta L \cdot \frac{N_s}{k_s} \right] + a \cdot b \cdot k_b \cdot N_b \tag{6.14}$$

Working load of the shaft group, Q_{ag} is

$$Q_{ag} = \frac{Q_{ug}}{2.5} \quad \text{(Factor of safety} = 2.5)$$

where ΔL = thickness of soil stratum; N_s = SPT value of stratum with thickness ΔL; N_b = SPT value at the base of block; k_s = empirical design factor for shaft friction; k_b = empirical design factor for end bearing; and a & b = length and breadth of the group in plan (in ft.)(see Figure 6.3).

Chan (2004) reported typical values of $k_s = 23-32$, and $k_b = 0.3-0.5$ for the case of dense cemented clayey sand with SPT $N \geq 50$. For loose to medium dense sand with SPT N of 5–30, Meyerhof (1976) recommended that $k_s = 100$. These factors can be verified/measured from a full-scale load test on instrumented test piles.

The ultimate carrying capacity of the shaft group may also be estimated using the modified Meyerhoff method as shown below. This method is also widely used locally.

$$Q_u = \sum \left(9.81 N_s \cdot \frac{A_s}{k_s} + 9 \cdot 81 k_b \cdot N_b \cdot A_b \right) \tag{6.15}$$

where Q_u = ultimate axial load capacity of the shaft (kN); N_s = SPT value along the shaft; A_s = perimeter area of the shaft (m^2); k_s = empirical design factor for skin friction; k_b = empirical design factor for end bearing; N_b = SPT value at the base; and A_b = cross-sectional area of the base. The value of k_s ranges from 2.5–3 (m^2/kN) and k_b ranges from 3–4 (kN/m^2).

6.3   CONSTRUCTION OF A DRILLED SHAFT

In firm soils, the dry method (also known as the open hole method) can be used to build the shaft. The construction procedure is generally as follows:

a. A cylindrical hole is excavated into the ground at the predetermined position of the drilled shaft, to the required depth using a drill rig equipped with a drill or bucket auger.
b. The lower portion of the shaft is then filled with concrete.
c. A prefabricated reinforcing steel cage is placed inside that shaft. Circular concrete spacers are attached to the cage to ensure that a minimum 60 mm concrete cover is provided.
d. The rest of the shaft is filled with concrete. For the dry method, ready mixed concrete with a slump of 100 to 150 mm is normally used.

The construction sequences are illustrated in Figure 6.4. Figure 6.5 shows a drill rig.

During drilling of the shaft, certain construction control measures can be implemented such as checking for verticality, *in situ* soil sampling and final determination of the founding depth of each shaft.

In the case of caving or squeezing soil, that is, where the hole will experience side collapse before or during placement of the concrete, two most common construction techniques for preventing these problems are the use of casing or the use of drilling fluid.

In the case of the casing method, the following procedures are generally followed (see illustration in Figure 6.6):

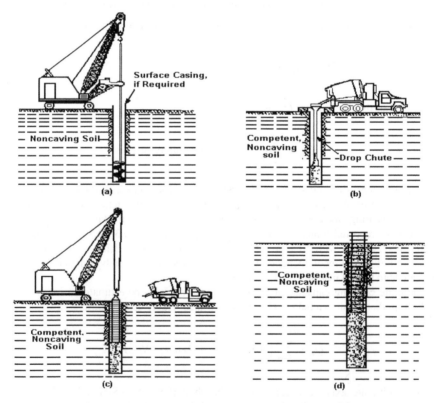

Figure 6.4.   Drilled shaft construction in competent soils using dry method. (a) Drilling the shaft; (b) Starting to place the concrete; (c) Placing the reinforcing steel cage; (d) Finishing the concrete placement (after Reese & O'Neill 1988).

Figure 6.5.   Photograph of a drill rig showing an auger drill at the tip.

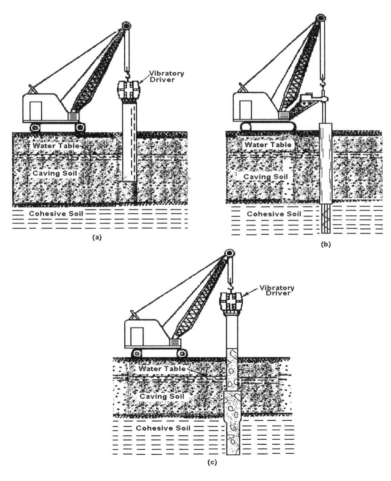

Figure 6.6.   Using casing to deal with caving or squeezing soils. (a) Installing the casing; (b) Drilling through and ahead of the casing; (c) Placing the reinforcing steel and concrete, and removing the casing (after Reese and O'Neill 1988).

Figure 6.7.   Photograph showing the steel casing in the background. In the foreground is the drill rig fitted
with a cleaning bucket.

  i. The hole is drilled in conventional methods until encountering the caving strata.
 ii. A steel pipe (casing) is inserted into the hole and advanced past the caving strata. Usually
     vibratory hammers are used for this purpose. The diameter of this casing is less than the
     diameter of the upper part of the shaft.
iii. Drilling is done through the casing and into the non-caving soils below using a smaller diameter
     auger.
 iv. The reinforcing steel cage and the concrete are placed through the casing, and the casing is
     extracted simultaneously using a vibro extractor. This is the most critical step, because the
     premature extraction of the casing can produce soil inclusions in the shaft.

There are, however, several variations to this method, including the option of leaving the casing
in place and combining the casing and slurry methods.

In cases where the hole cannot be kept dry, the hole should be filled with water up to at least the
depth of the ground water level. In this case the tremie method of concreting can be adopted with
very workable high slump (150–200 mm) concrete.

The drilling fluid method (also known as slurry method) is illustrated in Figure 6.8. It uses the
following procedure:

  i. A starter hole, approximately 3 m deep is drilled.
 ii. The starter hole is filled with a mixture of water and bentonite clay to form a drilling mud or
     slurry. This material must have a consistency to keep the hole open because of the hydrostatic
     pressure it applies to the soil.
iii. The hole is advanced passing the drilling tools through the slurry. At the same time, adding of
     necessary amounts of water and bentonite is continued.
 iv. The reinforcing steel cage is inserted directly into the slurry.
  v. The hole is filled with (high slump) concrete using a tremie pipe that extends to the bottom of
     the hole. The concrete pushes the slurry to the ground surface, where it is captured.

This method is quite effective but slow and messy.

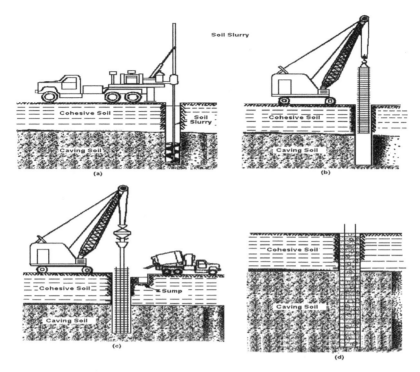

Figure 6.8.   Using drilling fluid to deal with caving or squeezing soils. (a) Drilling the hole using slurry; (b) Installing the reinforcing steel cage through the slurry; (c) Placing the concrete using a tremie pipe and recovering slurry at the top; (d) The completed foundation (after Reese & O'Neill 1988).

6.4   INSTRUMENTED DRILLED SHAFT

Liew et al. (2004) describe the results of an instrumented drilled shaft at a site located at Bukit Jalil, Kuala Lumpur, Malaysia. The site is underlain by weathered meta-sedimentary soils of the Kenny Hill formation, which consists mainly of sandy clay and silty sand (Figure 6.9).

The instrumented drilled shaft was 1 m in diameter, 20 m long, designed as a friction pile with a working load of 3600 kN. Figure 6.10 shows the instrumentation details of the drilled shaft, which comprise five levels of strain gauges and four extensometers. The drilled hole was wet when reaching the lower portion of the shaft. An overbreak at depths from 1.5 m to 7 m below the ground level was also observed. Maintained static load test was performed on the instrumented drilled shaft eleven days after the installation. This shaft was only loaded and unloaded in two cycles with test loads of up to 7200 kN (i.e. twice the design load). The applied load was maintained for 15 minutes for each load increment and was maintained for 6 hours at the working load, and twice the working load.

Load transfer curves for the shaft resistance and base resistance are shown in Figure 6.11. The ultimate shaft and base resistance was apparently not fully mobilized, as the load carried along the shaft continued to increase with increasing pile settlement during the load test. Higher shaft resistance was observed in the upper section with an overbreak recorded, due to the bulged section. Soft toe condition was evident as shown in Figure 6.11, in which the base resistance only started to be mobilized after about 1 mm settlement had occurred at the base. Liew et al. (2004) suggested that base resistance should ideally not be considered in a drilled shaft design, unless base cleaning for proper base contact can be carried out.

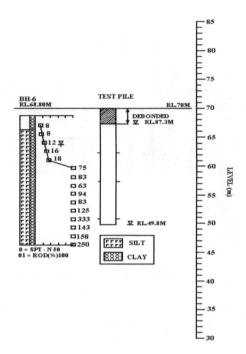

Figure 6.9.   Borehole logs of the test site.

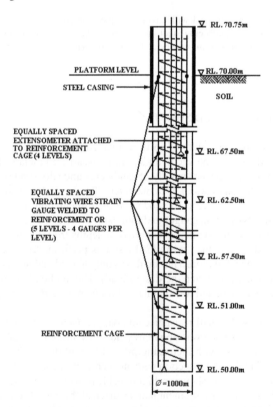

Figure 6.10.   Details of the instrumented drilled shaft (from Liew et al. 2004).

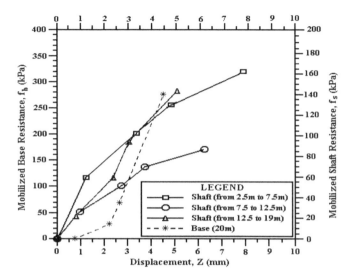

Figure 6.11.   Load transfer curves of the instrumented drilled shaft test pile (after Liew et al. 2004).

Figure 6.12.   The Petronas Twin Tower, Kuala Lumpur, Malaysia.

6.5   BARRETTE PILES

Where high horizontal load and moment capacity forces together with axial forces are anticipated and where regular pile sections fail, the usage of barrette pile prevails. The borehole for the drilled shaft can be excavated by percussion to make excavations with non-circular cross sections. A surface casing, or guide, in the form of a cross or a rectangle can be placed. The transverse dimensions of the guide will conform to the size of the grab bucket. A drilled shaft of this type is called a "barrette". These barrette piles have been used for the foundation of the Petronas Twin Tower (Figure 6.12) in Kuala Lumpur, Malaysia. The foundation under each tower is a 4.5 meter thick, 32,500 tons raft containing 13,200 m^3 of grade 60 concrete, and supported by 104 barrette piles, 30 to 108 m deep, driven into the soft limestone rock.

REFERENCES

Burland, J.B. 1983. Shaft friction of piles in clay – a simple fundamental approach. *Ground Engineering.* 6(3): 30–42.

Chan, S.F. 2004. Design and construction of foundations for Suntec City, Singapore. *Malaysian Geotechnical Conference.* The Institution of Engineers, Malaysia. Kuala Lumpur. 21–43.

Kulhawy, F.H. & Jackson, C.S. 1989. Some observations on undrained side resistance of drilled shafts. Foundation Engineering: Current Principles and Practises. ASCE. 1011–1025.

Liew, S.S., Kwong Y.M. & Gan S.J. 2004. Interpretation of instrumented bored piles in Kenney Hill Formation. *Proceedings of Malaysia Geotechnical Conference.* The Institution of Engineers Malaysia. Kuala Lumpur. 291–298.

Meyerhof, G.G. 1976. Bearing capacity and settlement of pile foundations. *Journal of Geotechnical Engineering Division.* ASCE. 102(3): 197–228.

O'Neill, M.W. & Reese, L.C. 1999. Drilled shafts: construction procedures and design methods. Federal Highway Administration. USA.

Reese, L.C. & O'Neill, M.W. 1988. Drilled shafts: construction procedures and design method. FHWA-HI-88-082. Federal Highway Administration. USA.

CHAPTER 7

Caisson and Well Foundation

Yee Yew Weng
Keller (M) Sdn. Bhd, Kuala Lumpur, Malaysia

7.1 INTRODUCTION

The hand-dug caisson is, as its name suggests, constructed primarily by hand excavation. Temporary cast in situ concrete linings, as illustrated in Figure 7.1, are constructed to retain the soil as excavation progresses. When the excavated shaft has reached the founding strata, usually bedrock, the shaft is filled with reinforced concrete. Caisson carrying capacity is derived principally from end bearing.

The hand-dug caisson is technically viable as long as the soil can be left free-standing for the lift height during excavation and the rate of water inflow is low. For the majority of residual soils in Malaysia, this criterion can be satisfied. Application of hand dug caissons can be found mainly on hill slopes and congested sites where conventional piling rigs have accessibility problems.

The method was originally developed in Hong Kong (Mak 1993), although health issues related to workers have resulted in their decline. Indeed safety and health issues are the main concerns with this labor intensive foundation construction method. Designers should give these issues due thought before specifying this foundation system.

7.2 APPLICATION OF CAISSON AND WELL FOUNDATION DESIGNS

The hand-dug caisson thrives in an environment where conventional piling equipment struggles e.g. hill sites and congested sites. Being of relatively large diameter (to be able to fit a worker therein), the caisson has potentially large structural capacity – typically above 500 ton. Where the rockhead is near the ground surface, say less than 20 m, the caisson can be socketted into rock to derive full structural capacity. Caisson construction works well on hill slopes and in residual soil

Figure 7.1. In-situ concrete linings of a hand-dug caisson.

where the groundwater table normally does not pose a concern but elsewhere where the water table is high, some form of groundwater control is normally required.

In Malaysia, the method has been used in a variety of projects in many different soil types. It has been found to be particularly useful under the following circumstances:-

Foundation

(1) Hill site development
Caisson foundation has been proven to be an effective foundation solution in hill sites where a combination of site access, limited space and safety make conventional piling methods impractical. Caissons can be designed with large diameters and each caisson can support highly loaded single columns. From the author's knowledge, the highest density of usage of hand-dug caissons can be found in Genting Highlands, Pahang, Malaysia. Due to scarcity of flat land, most of the high-rise buildings there are constructed on sloping ground. Conventional piling would not only have required expensive platform and access road construction but would have damaged large tracts of neighbouring forest land. The caissons were generally founded deep socketted into granite rock. On a similar note, many bridges and road widening projects carried out on the hill resort employed hand-dug caisson as the foundation scheme.

(2) Mixed ground conditions
Built ground or ground filled with boulders provide the foundation engineer with many challenges. Driven piles may stop short on localized gravel layers. It may cost a lot to drill pass boulders. Verifying the founding layer is generally not easy. Hand-dug caisson has been proven to be effective in such grounds.

(3) Environment constraints
Caisson construction requires a headroom clearance of less than 3 m (for lifting the bucket). It is generally a quiet process except where breaking through rock. Hence, where foundations are required in areas of low headroom (e.g. under an existing bridge) or where noise restrictions exist (e.g. hospital), hand-dug caissons can be considered.

(4) Difficult ground conditions
The hand-dug caisson has been used in very difficult ground conditions including in cavernous limestone. For example, the 40 storey Menara Kek Seng (Jalan Bukit Bintang, Kuala Lumpur) was founded on 4 m diameter caissons socketted into limestone. The integrity of the founding rock was probed within the caisson and cavity treatment instituted (Yee & Yap 1998).

Retaining wall

(1) Caisson retaining wall
Large diameter caissons have been used extensively in hill sites in Genting Highlands, Cameron Highlands, Penang and Kuala Lumpur, Malaysia. One of the biggest projects in Malaysia involving such vertical walls was constructed in Genting Highlands, involving a two-tier 40 m high retention system (Wallace & Yee 1996). Figure 7.2 shows an example of a caisson retaining wall.

(2) Landslide protection barrier wall
Barrier walls against boulder slides have been constructed using caissons on hillsites.

Top-down construction

In terms of technology advancements, the hand-dug caisson has little to offer. However, some innovative design and construction methods have been made possible with the help of this technique. Among them is the Menara Prudential Tower (Jalan Sultan Ismail, Kuala Lumpur, Malaysia) where 4-m diameter caissons support the 20-storey tower on shaft friction in Kenny Hill Formation.

Figure 7.2. A completed caisson retaining wall complete with fencing installed at the top of the capping beam (photo courtesy of Kumpulan Ikram Sdn. Bhd.).

The site was a former petrol kiosk and space availability was so tight that top down construction was specified. Construction time was shortened significantly when a permanent concrete column was constructed within the caisson shaft to carry the tower load as the basement was being excavated at the same time.

Investigation work

Construction of a hand-dug caisson is essentially a well digging exercise. During the excavation of caisson wells, the soil conditions can be assessed and buried objects can be examined. Such wells have been used in forensic work of failed structures where the soil conditions and pile conditions (depth and quality) can be examined (Yee 2003).

7.3 DESIGN CONSIDERATIONS

Soil investigation

As in the case of all foundation work, adequate soil investigation is a prerequisite for efficient design. Sufficient soil data is also important to assess cost of construction and minimize unforeseen risks. The objectives of the soil investigation should include the following:

  i. *Determine soil type.* Understand the ground conditions that will be encountered during excavation. The hand-dug caisson is only viable as long as the soil can be left free-standing for an unsupported height during excavation (typically 1 m) and the rate of water inflow is low. Very soft clays or slime would be unsuitable for caisson construction. Soil parameters are derived for lateral pressure assessment and caisson ring design.
 ii. *Determine the founding layer and level for the caisson.* Assess the depth of excavation required and corresponding practical issues.
iii. *Determine the strength characteristics of the founding rock.* Assess founding soil/ rock strength through in situ SPT test or retrieve samples for examination and/or perform UCS test.
 iv. *Determine the groundwater level.* This is a very important exercise as excessive groundwater inflow is a real concern for caisson construction. Tests should be done to predict groundwater inflow characteristics to ensure construction work can proceed smoothly and to check if groundwater drawdown will affect neighboring properties. Water pumping tests may be necessary. Water data is also required for caisson ring design.

It is common to sink one borehole beneath every caisson, especially where the foundation is expected to carry large loads. In residual soil and rock, it is important that a geologist be employed to ensure that the rock/soil gradings (Grades I to VI) are properly identified before specifying the founding depth.

Caisson Ring

The caisson ring design is not difficult but its importance cannot be over emphasized. The ring is defined as temporary work and design is usually left to the contractor. Failure of the ring can have dire consequences in terms of safety and program delay. The following points should be considered in the design of the caisson ring.

i. *The concrete ring is constructed* in situ. In most cases, the concrete is not vibrated. It would have to gain strength quickly to support the vertical soil cut as the steel mould is normally removed within 24 hrs.
ii. *The reinforcement has to be placed in very tight working conditions.* Nominal compression steel is normally provided (e.g. steel mesh A4, A6, A10). Additional steel may be required in localized soft soil layers e.g. tension anchorage bars should be provided where there is soil instability concern. In some cases, heavier reinforcements are required to cater for surcharge loading or external lateral forces.
iii. The depth of excavation, soil behavior, water inflow, vibration levels from surrounding works, etc. would have to be assessed. Needless to say, the ring should have to be designed to remain intact under all construction conditions.

The depth of each lift (unsupported soil dig) is determined mostly by experience to ensure that the soil would not collapse. For example, in dry soil, depths up to 1.2 m can be constructed. In wet soil, the dig may need to be reduced to 0.75 m. For large diameter caissons, the depth of each lift may have to be reduced as restricted by the weight of the steel mould.

The box below outlines step for the design of a caisson ring.

Design of Caisson Ring

1. $\sigma_r = \sigma_h r / t$
σ_r = stress on concrete ring
σ_h = horizontal earth pressure
r = radius of concrete ring
t = thickness of caisson ring

2. Check $\sigma_r < \alpha f_{cu}$
α = reduction factor for concrete strength achieved in 24 hrs
f_{cu} = concrete strength in 28 days

3. Provide reinforcement
i. nominal steel in compression (BS8110 Table 3.27)
ii. tension anchorage (BS8110 3.12.8.4)

Water inflow

One of the most important considerations in caisson design lies in the assessment of ground-water inflow. Experience of the engineer, in this regard provides the most important guide. Data on groundwater table should be carefully studied, especially to confirm if there is any seasonal variation (e.g. due to rain), tidal fluctuations and aquifer pressures. Soil investigation should include installation of piezometers which are monitored over a certain period. Water fluctuation on soil slopes can be assessed by installation of Halcrow buckets.

Soil samples should be assessed carefully, especially where the water table is near ground level. Uncontrollable water inflow through sandy layers (even very dense SPT >50 blows/ft stratum), have resulted in many hand-dug caisson excavation work being abandoned. Where there is doubt, the designer should perform some pumping tests to assess permeability at the depth concerned. The risk of piping or basal heave should be studied. Trial digging of a single well may be done where design is uncertain.

Dewatering during caisson construction would lead to lowering of the groundwater table. This may result in subsidence of the surrounding ground, services, roads or properties. The risk of such occurrences should be studied carefully as property damage cases have been reported in some projects.

As a very general rule-of-thumb, deep caissons (say above 10 m) in water bearing ground with sand content above 50% would require careful consideration. The safety of workers should not be compromised and an alternative foundation system should be explored where there is doubt. Alternatively, pre-excavation soil treatment may be instituted. This can be in the form of curtain grouting or jet grouting at the permeable stratum.

Axial capacity

Caisson carrying capacity is normally derived from the rock socket. The contribution is predominantly due to base bearing although rock friction can form a significant proportion of the carrying capacity for smaller caissons. Caissons founded in dense soil are becoming more common although there are design issues related to reliability of shaft friction.

(a) Shaft friction

Typical rock friction values used for pile design can be found in many publications, such as Tomlinson (1977), Horvath and Kenney (1979) and Toh et al. (1989). Socket friction should be ignored where the caisson is constructed "bell-out". In reality, due to the method of construction of using hand chisels, most caissons are constructed with a slight bell-out. Shaft friction in soil is normally neglected due to uncertainty in the quality of bonding of the lining with the surrounding soil and with the caisson shaft. Limited tests done locally have shown that load transfer can be effective where careful construction is carried out (Yee 2001).

Some authoritative publications recommend using an empirical correlation to derive rock friction based on rock unconfined compressive strength (UCS) tests. Recent work in Hong Kong has proposed a design relationship for a whole range of rock types and strengths. However, it should be noted that UCS test results may vary significantly even for samples recovered over short distances. The results may not be consistent with the degree of weathering. Further, correlations involving UCS may only be useful for stronger rock where intact rock samples can be recovered. For highly weathered or highly fractured (or weathered) rocks, it may not be possible to recover intact samples for testing. For these rocks, it may be possible to rely on empirical correlations based on SPT blowcounts "N". This type of design approach is widely used in Malaysia for pile design in completely weathered sedimentary rock (Tan et al. 1998). Wherever possible, full scale load test should be carried out to verify design assumptions.

For caissons embedded in slopes, some designers advocate the use of friction reducing sleeves to prevent any transfer of load due to possible slope movement. Alternatively, the anticipated dynamic load can be designed to be resisted by the caisson. Caisson performance may be modelled using Viggiani's (1981) method.

(b) End bearing

Typical rock bearing values used for caisson design are generally derived from work done for drilled shafts e.g. Tomlinson (1995 and 1997), Poulos and Davis (1980), HK GEO (1996), Hill et al. (2000). As can be expected, suggested bearing values vary significantly probably because experience from different authors were derived from different rock type at varying degrees of grading/weathering. As for rock friction, empirical correlations to derive end bearing using UCS and SPT "N" values are widely used among designers. Unlike other pile types, the founding rock for hand-dug caissons may be carefully cleaned by the contractor and subsequently examined visually by the engineer. In situ probing tests could be carried out on the base to ensure that there is no softer layer beneath the founding level. Hence, the reliability of the end bearing component of the caisson compared to bored pile is normally expected to be higher.

Lateral load capacity

Caissons may be required to carry lateral loads from bridges, retaining walls or to cater for landslide events. The lateral capacity of a caisson is generally governed by:

1. soil shear capacity;
2. pile structural capacity; and
3. excessive pile deflection

Soil investigation should be carried out to obtain design parameters such as soil modulus E and strength parameters c' and ϕ'. The expected pile performance may be predicted using Brinch Hansen's method (cited in Tomlinson (1977)) given by the formula:

$$P_{ult} = (k_1 \cdot K_q \cdot \sigma'_v + k_2 \cdot K_c \cdot c') \cdot h \cdot D \, (\text{kN})$$

$$K = E_s \cdot h \, (\text{kN/m}),$$

where
k_1, k_2 = correction factors to suit different ground condition; K_q = Brinch Hansen's coefficient for frictional component; K_c = Brinch Hansen's coefficient for cohesive component; h = the distance between the mid point of elements above and below the node and D = the pile diameter.

Computer programs are available for modeling pile behavior. The empirical correlation of $E = 1$ 'N' (N = SPT blowcount) has been commonly used to model soil behavior among designers. Recent load tests by Plumbridge et al. (2000) have shown that this may be conservative as E (MPa) between $3N$ and $6N$ were derived.

Group effects

Caissons normally derive their carrying capacity through end bearing. BS8004 specifies that the clear spacing between end bearing piles should not be lower than one pile diameter (shaft to shaft). HK GEO (1990) allows minimum clear spacing of 0.5 m between shaft surface. The minimum clear spacing should also be based on practical considerations of construction tolerance. Also, where there is significant variation in rock bearing levels, it is sometimes prudent to increase the depth of one caisson to ensure that is does not transfer load onto an adjacent caisson.

Where a caisson relies on shaft friction, the shaft to shaft spacing should not be less than two times the caisson diameter (BS8004). If this is not physically possible, then the pile capacity may have to be downgraded appropriately.

Settlement

The carrying capacity of a caisson socketed into intact rock is seldom governed by settlement. For caisson founded on weathered rock, the settlement at the surface of the rock mass may be calculated by the following formula:

$$\delta_b = \frac{Q(1 - v_r^2)D_b}{E_m} C_d C_s$$

where $\delta_b =$ settlement at the surface of the rock mass; $Q =$ bearing pressure on the rock mass; $C_d =$ depth correction factor; $C_s =$ shape and rigidity correction factor; $v_r =$ poisson's ratio of the rock mass; $D_b =$ pile base diameter and $E_m =$ modulus of rock mass.

Back-analysis of load tests on bored piles socketed into granite by Hope et al. (2000), found E_m between 0.3 MPa and 5.1 MPa.

7.4   CAISSON LOAD TESTS

Full scale tests are generally expensive due to the high test loads required and since caissons are generally constructed in difficult terrain. The heavy kentledge weight required may also affect test results non-conservatively since the high vertical stresses transferred onto the ground may stiffen the test caisson response. In general, tests are done on scaled down models (e.g. shorter and smaller diameter caissons or piles) and the actual caisson behavior extrapolated. This is not an unreasonable approach provided that the load transfer mechanism is modeled appropriately. For example, tests to evaluate end bearing capacity should firstly, take into account the magnitude of load shed through shaft friction and compensate for it appropriately.

Data from a few full scale load tests under different loading and ground conditions have been made available to the author and are discussed below.

(a) Load test in granite rock

Many pile tests in rock do not distinguish between the end bearing and shaft friction load carrying components. Strain gauges can be embedded in the piles to study load distribution but in most cases, much of the load is carried in shaft friction and very small loads are transferred to the base. Effective debonding of the shaft is difficult to achieve and expensive means such as double casing have to be specified. Yee et al. (2001) reported full scale load tests conducted in Malaysia using pressure cells cast into the caisson at various levels to enable modeling of load transfer for both side friction and end bearing components.

Based on the limited data available, there seems to be an empirical relationship between shaft friction and UCS of rock samples collected from the bearing rock. For completely weathered granite, the commonly used relationship $f_{su} = 2N$ seems to be reasonable. The tests proved that the caisson ring could transfer load onto the soil provided construction is carried out carefully.

The end bearing capacity of the decomposed granite soil is relatively high ($f_{bu} = 60N$) although this was generally accompanied by some settlement. The modulus E was determined to be between 0.2 and 3.7 MPa for intact rock; and 0.04 MPa for the Grade V completely decomposed granite. This is consistent with published data.

(b) Load test in Kenny Hill Formation

Limited plate load tests have been carried out within the base of an excavated caisson (Arup 1996 to 2001). The end bearing mobilized was relatively high ($f_{bu} = 50N$). The modulus E was found to be 70,000 kPa.

Shaft friction was evaluated by placing a series of jacks within a gap between constructed caisson rings and applying load to push them apart. This simple test showed that the caisson ring could transfer load onto the soil. The commonly used relationship $f_{su} = 2N$ seem to be acceptable.

(c) Full scale lateral load test

Very few lateral load tests results are available to the author. One lateral load test was carried out at a hill-site on a 1.2 m diameter caisson embedded on sloping ground. The applied lateral load of 80 ton caused a maximum 30 mm deflection. Back-analysis of the data indicate a relationship of $E = 1500N$ and $k_1 = 0.5$.

7.5   CONSTRUCTION ISSUES

Sequence of construction

The construction of a hand-dug caisson involves many days and sometimes runs into weeks depending on the depth and hardness of the soil to be removed. As such, it is normally necessary for the contractor to plan and design the work to suit the conditions. Typical steps in construction are:

a. Prepare platform for caisson work.
b. Set-up hoist system. This consists of an electrically operated hoisting system supported by a frame which consists of steel runway beam and supports.
c. Excavate vertically downward in lifts of between 750 mm and 1200 mm to form a circular shaft.
d. Lower steel mould and construct concrete lining (ring) in situ. The mould is left in place for 24 hrs while the concrete ring cures to support the earth pressure.
e. Excavate the next lift and construct concrete lining.
f. Repeat progressively until excavation has reached the founding rock stratum.
g. Excavate required rock socket using chisel and rock splitting methods. Explosives are sometimes used but this is not recommended in a confined environment. Concrete lining is normally not required for excavation of the rock socket.
h. Confirm rock socket suitability by geologist or suitably qualified personnel. Compare with soil investigation data.
i. Perform in situ test by drilling with hand held tool (typically 5 m below base).
j. Inspection by experienced engineer and approval for construction.
k. Lower pre-fabricated steel cage and pour concrete by tremie method.

Termination criteria

Termination criteria of the caisson are usually specified by the engineer in the drawings. However, in many cases, because the caisson normally will carry high loads and its performance is critical, inspection of every caisson by the engineer is recommended. This may involve three levels of check:

1. Ensure that the excavation level has been reached as shown in the drawings (as determined from soil investigation data and design calculations).
2. Examination of rock quality to verify quality and consistency throughout the socket length.
3. Review in situ test results e.g. by carrying out rock drilling tests (check rate of penetration) or localized plate load tests.

Although this foundation type has the advantage of enabling visual examination, cases of misinterpretation of rock founding layer have been known to occur. For example, core boulders in granitic soils can be easily misinterpreted to be Grade II bedrock. Hence, an experienced geotechnical engineer or geologist should be responsible for examining and verifying every caisson bearing stratum. Where there is doubt, additional soil investigation should be carried out to confirm the integrity of the founding layer.

Construction time

The construction of many caissons can proceed simultaneously on a site because it does not rely on the availability of piling rigs; is not governed by availability of material or space constraints. Hence, construction can be relatively quick provided that labor resources are available in abundance. Based on past experience, caisson work down to a reasonable depth would not be slower than conventional piling works. Depending on the soil conditions and diameter of the caisson, a dig of 0.75 m to 1.2 m can be constructed each day. Typically, a 1.5 m diameter caisson about 20 m deep would normally take about 3 weeks to construct.

Figure 7.3.   Hand dug caissons for a hill site development.

Construction cost

Based on general comparison of rates, the cost of hand-dug caissons are normally higher than conventional piling. Hand-dug caissons usually become economically viable compared to other conventional piling under unconventional circumstance such as:

- difficult site conditions e.g. hill-site (Figure 7.3) or congested site
- mixed ground conditions e.g. boulders
- large capacity single load carrying element required.

7.6   SAFETY CONSIDERATIONS

Hand-dug caisson construction involves workers working in deep shafts subjected to risks. Common causes of accidents involve falling objects, ingress of water and persons falling down the shaft. Health risks would include dust inhalation and noise damage hearing. Some of the safety precautions recommended would include:

1. Construct barricade at top of caisson ring to prevent objects from falling into hole.
2. Ensure workers are working in pairs and that they are equipped with helmet, whistle, walkie talkie, workshoe, ear plugs and breathing mask.
3. Provide adequate lighting and fresh air supply at all times.
4. Double chain hoist should be provided and regularly maintained.
5. Workers should stand under a protective temporary shed during lowering or lifting of hoist bucket.
6. Excavated soil should be disposed off quickly to avoid surcharging the caisson ring.

Despite the best precautions, long term health risks cannot be eliminated. Exposure to the daily operation of vibrating chiseling tools and inhalation of fine particulates will have long term health effects. Designers should give this matter due consideration before specifying this foundation system.

7.7 SUMMARY

The hand-dug caisson or well foundation system has found wide application in Malaysia. The design of the caisson is rather straightforward but the engineer is required to have a proper understanding of the ground conditions, together with the founding rock characteristics. In particular, the constructability of the caisson in the soil and groundwater conditions must be given careful consideration. Available full scale caisson test data is scarce because of the relatively high cost for conducting such tests and since most applications are in steep and congested sites. Wherever possible, foundation caissons should be load tested. Some design rock parameters have been suggested in this chapter. These have been mainly derived using empirical methods (e.g. correlation with UCS tests) and limited load tests results.

Although the construction time for a single caisson takes a relatively long time, the overall foundation construction program may not be slower compared to conventional piling provided that the caissons can be opened up all at once and labor is available in abundance. Caisson construction is relatively expensive and normally becomes viable in difficult ground or site conditions, where other types of piling methods become costly.

High safety standards must be maintained during construction. Detrimental long term effects on the health of caisson construction workers cannot be eliminated and hence, it is recommended that the usage of such hand-dug methods should be limited to developments where other methods are impractical.

REFERENCES

Arup Jururunding Sdn Bhd. (1996 to 2001). Test pile reports. Private Communications.
BS8004. 1986. Code of practice. Foundations. BSI.
Geo Publication No. 1/96. 1996. Pile design and construction. Hong Kong.
Hill, S.J. et al. 2000. End bearing and socket design for foundations in Hong Kong. *Proc. 19th Geotech. Seminar (HKIE) – Foundations*. Hong Kong.
HK GEO 1990. Foundation design building (Construction) regulations – Part VI. PNAP 141.
Hope, S., Young, S. & Dauncey, P. 2000. Airport railway pile tests. *Proc. 19th Geotech. Seminar (HKIE) – Foundations*. Hong Kong.
Horvath, R.G. & Kenney, T.C. (1979). Shaft resistance and rock-socketed drilled piers. *Symposium on Deep Foundations*. ASCE: 182–214.
Mak Y.W. 1993. Hand-dug caissons in Hong Kong. *The Structural Engineer*. 71(11).
Plumbrige, G.D., Sze, J. & Tham, T. 2000. Full scale lateral load tests on bored piles and a barrette. *Proc. 19th Geotech. Seminar (HKIE) – Foundations*. Hong Kong.
Poulos, H.G. & Davis, E.H. 1980. *Pile foundation analysis and design*. John Wiley & Sons, Inc.
Tan, Y.C., Chen, C.S. & Liew, S.S. 1998. Load transfer behavior of cast-in-place bored piles in tropical residual soils of Malaysia. *Proceedings 13th SEAGC*. Taipei.
Toh, C.T., Ooi, T.A., Chiu, H.K., Chee, S.K. & Ting, W.H. 1989. Design parameters for bored piles in a weathered sedimentary formation. *Proceedings of 12th International Conference on Soil Mechanics and Foundation Engineering*, Rio de Janeiro. 2: 1073–1078.
Tomlinson, M.J. 1977. Pile design and construction practice. Viewpoint Publication.
Tomlinson, M.J. 1995. *Foundation design & construction*. 6th Edition. Longman.
Viggiani, C. 1981. Ultimate lateral load on piles used to stabilise landslides. *Proc. 10th ICSMFE*. Stockholm.
Wallace, J.C. & Yee, Y.W. 1996. Site stabilisation and foundations for a high-rise building on a steep slope. *Proceedings 12th SEAGC*, Kuala Lumpur. 1: 347–354.
Yee Y.W. & Yap L.C. 1998. Hand-dug caisson foundations for tall building over cavernours limestone. *Proceedings 13th SEAGC*. Taipei.
Yee Y.W., Pan J. & Guo Y.Y. 2001. Construction and design of hand-dug caisson in Malaysia. *Proceedings 14th SEAGC*. Hong Kong.
Yee Y.W. 2003. Case histories of piled foundation failure. *2 days Conference on Geotechnical Engineering: Design & Construction of Pile Foundation (The Malaysian experience)*. Kuala Lumpur.

CHAPTER 8

Special Topic: Design of Piles Subjected to Lateral Soil Movement

Tan Yang Kheng
Consulting Engineer, EK Consultants, Kuching, Sarawak, Malaysia

Ting Wen Hui
Consulting Engineer, Kuala Lumpur, Malaysia

8.1 INTRODUCTION

In a number of design situations, piles have to be designed for effect of lateral soil movement. These include piles in or near an embankment built on soft clay, bridge abutment piles in soft ground, piles adjacent to an excavation, piles in unstable slope and piles in a marginally stable riverbank. These piles are called passive piles (De Beer 1977). As these piles will experience additional stress and strain, failure to assess the effect in design will result in unacceptable pile movement or stress or both.

Cases of pile failure due to lateral soil movement have been reported by Tschebotarioff (1973), Chin (1979), Chan (1988), Tan (1988) and Ting & Tan (1997). In the IXth International Conference on Soil Mechanics and Foundation Engineering held in Tokyo in 1977, one specialty session was held to discuss the above problem. The specialty session is entitled "The Effect of Horizontal Loads on Piles due to Surcharge or Seismic Effects". This had resulted in greater awareness of the problem. Various methods of designing piles subjected to lateral soil movement were also discussed. Examples are presented by De Beer (1977), Marche & Scheneeberger (1977) and Franke (1977). Since then, more studies have been carried out. Many of these works have been either reported in international conferences or published in engineering journals or reports (Viggianni 1981, Springman & Bolton 1990, Stewart et al. 1994, Chow & Yong 1996, Chen & Poulos 1997).

In Malaysia, design of passive piles is still quite new to many practicing engineers. Problems of pile supports due to lateral soil movement, though not uncommon in Malaysia, are sometimes not recognized or not well understood. In view of the above, this Chapter is written to help the readers to understand the mechanistic behavior and design of passive piles. A good reference on the subject is Ting & Tan (2004) which proposes a viable methodology in the design and construction of piles in the environment of limiting bank slope stability in the light of the researchers' findings, which is augmented by some site observations.

8.2 PROBLEM IDENTIFICATION

There are a number of construction design situations in which piles will be subjected to lateral soil movement. These include:

1. Piles in or near embankment, like bridge abutment piles,
2. Piles in unstable slope,
3. Piles in marginally stable riverbank with high fluctuating water level, and
4. Piles near an excavation.

Tschebotarioff (1973) reported several cases of unfavourable pile supported bridge abutment movement in USA as early as 1950s. Chin (1979) cited an interesting case of pile failure of a building due to nearby surcharge in Malaysia. Chan (1988) noted that it is not uncommon to observe bridge

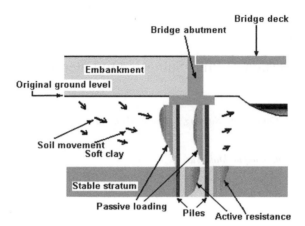

Figure 8.1.   Piles subjected to lateral soil movement due to nearby embankment.

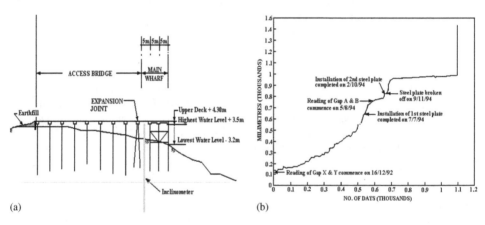

(a)   (b)

Figure 8.2.   (a) Cross section showing the riverbank and the wharf structure. (b) Gap measurement at the
expansion joint with time.

abutment failure due to unsymmetrical surcharge in soft clay areas in Malaysia. The problem is
described in Figure 8.1.

Ting & Tan (1997) reported another interesting case history of excessive movement of a riverside
structure subjected to high fluctuating water level. Findings from this case history are significant
because, in Malaysia, excessive structure movement of structures like wharves built in coastal
(invariably soft ground) areas are not uncommon, but mechanism of such failure has not been
identified before that.

Figure 8.2(a) & (b) briefly describe one such case for a concrete wharf built in a riverbank. Not
long after it was completed, the main wharf moved laterally at an average rate of 1.2 mm per day.
The movement is caused by lateral soil movement taking place in the riverbank and resulting lateral
loading induced on the piles.

8.3   FACTORS TO BE CONSIDERED

To design piles for lateral soil movement, the following factors need to be considered:

1. Soil movement mechanism.
2. Soil properties.

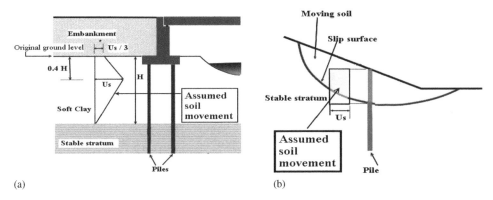

(a)

(b)

Figure 8.3.   (a) Assumed profile for soil movement due to nearby embankment. (b) Assumed profile for soil
movement in an unstable slope.

3. Pile properties.
4. Pile head condition.
5. Superstructure loading.
6. Ground support by lower stable stratum.

Soil movement mechanism

Soil movement is the cause of the problem. It will occur when there is unsymmetrical load applied
to the soil, or in an unstable slope, or loss of ground support during excavation. To analyse the
problem, profile and magnitude of the soil movement have to be defined.

For unsymmetrical loading (like embankment on soft ground), soil movement profile can be simplified as shown in Figure 8.3(a). In the case of unstable slope, the most critical slip surface will be
determined and a uniform soil movement to its full depth can be assumed as shown in Figure 8.3(b).

It would also be possible to apply other displacement profiles, provided that the assumed shape
is appropriate to the problem analysed. In practice reasonable displacement can be obtained from
theoretical considerations, past measurements of soil movement by inclinometers (Stewart et al.
1994), and by centrifuge models as in the work of Ong et al. (2003) in their study of excavation-
induced soil movement reproduced here as Figure 8.4.

Magnitude of soil movement (like U_s in Figure 8.3(a) & (b)) is a key aspect in design. However,
its value is difficult to estimate with reasonable confidence. Some general guidelines can be used
to evaluate U_s qualitatively. This is done with reference to factor of safety (*FOS*) against stability.
Discussion and suggested guidelines are described in Section 8.5.

Soil properties

Soil strength is another key parameter in design. Usually undrained shear strength, C_u, is used to
define soil strength. It is needed for determination of factor of safety (*FOS*) against stability, and
estimation of soil stiffness (E_s) and limiting soil pressure (y_s). Lower bound strength values may
need to be considered especially for sensitive soils subject to remoulding in the problem.

Some discussion on limiting soil pressure is needed. As soil is not an elastic material, it will
start to yield when a certain strain is reached. Thus pile-soil contact pressure will reach a limiting
value then. And ultimately loading on the pile will attain its peak value. This is quantified as force
per unit length of the pile, and is given by ($y_s \times$ pile width), where y_s is named as limiting soil
pressure. Its value is usually given as a function of C_u, i.e. $y_s = m \times C_u$, where m is a constant.
Different values of m have been suggested by researchers (Franke 1977, Viggiani 1981, Chen &
Poulos 1997, Ong et al. 2003). Methods of Hansen (1961) and Broms (1964) are commonly used
to determine limiting soil pressure. For conservative design, value of m can be taken as 7.5.

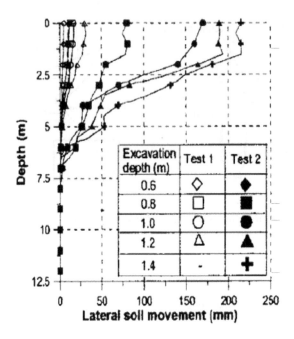

Figure 8.4.   Soil displacement profile (after Ong et al. 2003).

Pile properties

Relevant pile properties required for analysis of the problem are:

– Diameter or width,
– Length,
– Spacing between piles,
– Flexural stiffness (EI) (where E = Modulus of elasticity and I = Moment of inertia),
– Shear capacity, and
– Bending moment capacity or moment of resistance (M_r).

All the above parameters are required in design and can be defined with great certainty. Usually shear capacity is not a critical parameter. On the other hand, moment of resistance (M_r) can be critical. This is so because piles used in practice have a limit in bending capacity. When soil movement is sufficiently large, bending moment at certain point in the pile can reach its M_r value. Further soil movement will result in pile bending failure. This must be checked in design. Preferably, pile is designed so that induced maximum bending moment is less than its M_r.

Pile head condition

Response of pile subjected to lateral soil movement will vary with different pile head condition. Therefore pile head condition has to be assessed and defined in design. Pile head condition can be one of the following:

I.   Free in translation and rotation.
II. Free in translation but restrained in rotation.
III. Fixed in translation but free in rotation.
IV. Fixed in translation and restrained in rotation.

Pile responds differently under each of the above pile head conditions. These are illustrated in Figures 8.5 to 8.8. Results shown in the figures are related to a problem in which a long steel

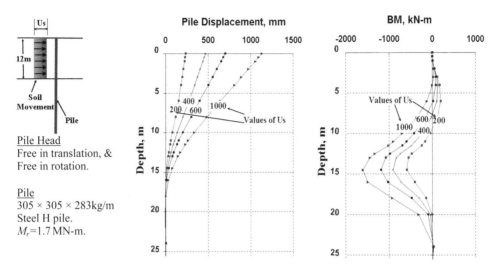

Figure 8.5.   Pile displacements and bending moment (BM) for steel pile with pile head condition I.

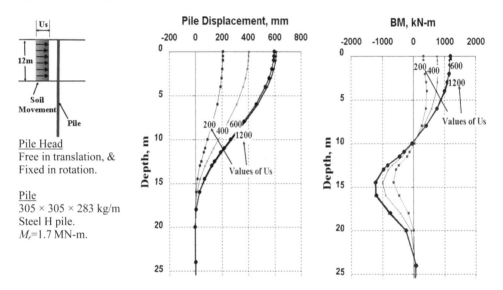

Figure 8.6.   Pile displacement and bending moment (BM) for steel pile with pile head condition II.

H pile ($305 \times 305 \times 283\,kg/m$) is subjected to a uniform soil movement to a depth of 12 m. All parameters of the problem are the same except its pile head condition. The problem was analysed using a computer software (ALP 1991). Computed pile displacement and bending moment (BM) are shown in the figures.

Figure 8.5 is for Case I in which the pile head is free to translate and rotate. The pile moves the most (when compared with other cases) and fails by bending when soil movement reaches about 1000 mm. In construction, this condition is likely if piles were not braced or tied with other adjacent piles prior to casting of pile caps.

Figure 8.6 is for Case II in which pile head is free to translate but restrained from rotating. It is observed that when soil movement reaches 600 mm, maximum loading has been mobilised. The pile displacement and bending moment reach the maximum. It is further observed that, for the steel

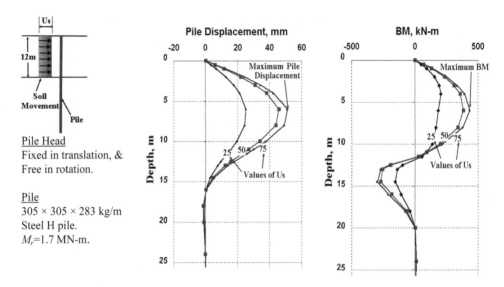

Figure 8.7.   Pile displacement and bending moment (BM) for steel pile with pile head condition III.

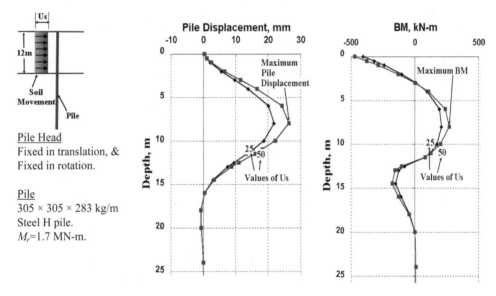

Figure 8.8.   Pile displacement and bending moment (BM) for steel pile with pile head condition IV.

H pile section chosen, the maximum bending moment is less than the pile moment of resistance. It is also worthwhile to compare the results with those in Case I. The only difference in the pile configuration is its pile head condition. In Case II, the pile head is restrained from rotating (which can be achieved partially on site by tying and bracing the pile head with adjacent piles). Beneficial effects of such restraining are obvious by comparing Figures 8.5 and 8.6.

Figure 8.7 is for Case III in which pile head is pinned and fixed in position. Maximum loading is reached at relatively small soil movement (about 75 mm only). Another observation is that maximum bending moment within the pile is also a lot smaller. The same trend is observed in Case IV, results of which are shown in Figure 8.8.

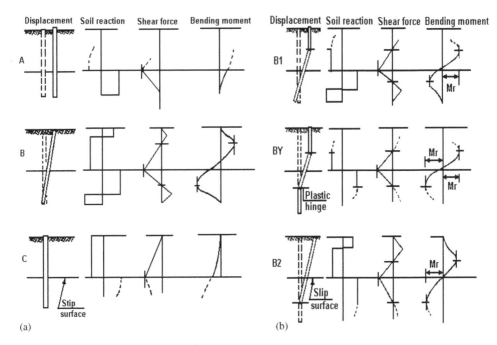

Figure 8.9.   (a) Modes of failure for rigid pile (after Viggiani 1981). (b) Modes of failure for flexible pile (after Viggiani 1981).

Superstructure loading

Superstructure loading will have an effect on the pile design when the imposed vertical and horizontal loads are large relative to the load solicited by lateral soil pressure on the piles.

Ground support by lower stable stratum

Only the stable stratum beneath the moving soil layer provides competent ground support to the piles. Therefore the pile length should be sufficiently long to satisfy the following conditions:

a. Failure mode is Mode C as described in Section 8.4.
b. Geotechnical capacity against axial load is adequate.
c. Pile head settlement is within acceptable limits under various loading conditions.

8.4   MODES OF FAILURE

It is useful to study different modes of failure of an isolated pile subjected to lateral soil movement. Figures 8.9(a) and 8.9(b) show different failure modes for rigid pile and flexible pile respectively. Rigid pile refers to pile with the moment of resistance that is more than the maximum bending moment induced. Flexible pile refers to pile with the moment of resistance equal to the induced bending moment at certain locations. In the latter case, plastic hinge(s) may be formed in the pile. Method to establish likely mode of failure is described by Viggiani (1981).

8.5   FACTOR OF SAFETY

One key aspect in designing piles subject to lateral soil movement is to evaluate factor of safety (*FOS*) against stability. Tschebotarioff (1973) has shown by correlating with case records that

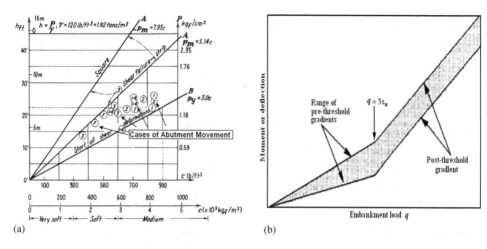

(a)   (b)

Figure 8.10.   (a) Plot of embankment height versus soil shear strength with cases of abutment movement
shown (after Tschebotarioff 1979). (b) Induced pile moment or deflection versus embankment
load (after Stewart et al. 1994).

Table 8.1.   Factor of safety and displacement.

FOS	1.0	1.20	1.50	1.70	2.00	3.00
Bending moment / Ult. Bending moment	1.0	0.65	0.38	0.28	0.19	0.08
Pile displ./Ult. Pile displacement.	1.0	0.64	0.37	0.27	0.18	0.07

substantial increase in bending of pile will commence when the subsoil is loaded to a certain level.
Figure 8.10(a) is reproduced from Tschebotarioff (1973). It shows that when embankment load
exceeded 3 C_u (corresponding to FOS = 1.7), significant movements were observed.

Stewart et al. (1994) has also arrived at the same conclusion by studying results of centrifuge
tests. Their results were summarized in Figure 8.10(b), reproduced from Stewart et al. (1994). It
can be seen that increase in induced pile moment and deflection increases significantly when the
embankment load exceeds 3 C_u (corresponding to FOS = 1.7).

Interpretation from the study of the results from Stewart et al. (1994) is summarized in Table
8.1. It can be seen that there is a distinct relationship between FOS and pile displacement; the
displacement becomes smaller as FOS is increased.

8.6   PILE DESIGN FOR LATERAL SOIL MOVEMENT

Criteria

Piles shall be designed to meet the following criteria:

1. Pile stability and structural resistance as constructed are to be determined as an isolated single
   pile and then as a pile incorporated in a group for:
   a. Moment and shear resistance at pile head and at slip surface and along the shaft of the pile.
   b. Axial capacity.
2. Adequate ground support to piles.

Certain possibilities of optimization of pile design exist before detail design is carried out. These
are briefly discussed here.

Table 8.2.   Lateral soil load on pile analysis.

Pile type	Diameter/size (m)	Moment of resistance (yield*) (kN-m)	Applied moment (kN-m)	Remarks
Conc. spun	1.0	1,053	2,400	140 mm thick
Tubular steel HT	0.6	1,928	1,463	299 kg/m
Steel H-HT	305 × 305	1,470	772	283 kg/m

* Cracking for concrete.

The lateral soil pressure on a pile is a function of the width of the pile. In the application of piles to resist moving ground, large diameter piles with high moment resistance are applied; with steel tubular piles often employed. In contrast, in the case of pile support required for the problem herein to minimize loading on the pile, a pile with minimum width and maximum moment resistance would be the optimum. Ting & Tan (2004) present results of three types of piles; reproduced here as Table 8.2.

It can be seen from the table that the steel H-pile has the best overall performance. It also has an advantage over the tubular steel pile in being the lower in displacement and thus causing lesser disturbance to the soft ground during pile driving. The concrete spun pile is unsuitable because of the low moment resistance.

The buckling of H-pile with respect to its weak axis in soft ground should be checked, especially in the case of partially embedded pile that has an unsupported length, which is often the case in bank problems. If required a built-up section at the upper part of pile may be provided. The enhancement of the upper section will also help to reduce dynamic instability (flutter) when driving long slender piles. The resulting deviation in pile alignment can be problematical in pile group with raking piles.

For pile groups the provision of raking piles will assist in reducing horizontal movement of the superstructure if it is an important consideration.

It was considered that with available knowledge (Ting & Tan 1997), "shielding" action of back rows of piles may be ignored and all individual piles in the group are subject to the limiting lateral soil pressure. The theoretical work of Chen & Poulos (1997) also showed that the assumption of full mobilization of limiting lateral soil pressure for all the piles in the group is not over-conservative for practical purposes. Their deduced "group factor" for a variety of cases has values ranging from 0.7 to 1.4.

Suggested design methodology

In designing piles subjected to large lateral soil movement, the following procedure is suggested:

Soil movement mechanism determination
To determine soil movement mechanism, carry out a stability analysis first. For slope stability analysis many programs are now available. The slope is analyzed for lower bound strength conditions and the degree of stability determined for all the potential slip surfaces. Usually, possible contribution towards stability by the embedded piles is not accounted for. The depth of lateral soil movement is deduced from the stability analyses. The appropriate lateral soil displacement profile to be applied on the pile within the depth of soil movement would be determined. Figures 8.3(a) and (b) provide some guidelines on probable soil movement profiles that can be adopted for design.

Evaluation of superstructure
Next design step is to study and evaluate the structure that is to be supported. Loading from the structure to the pile is quantified and will be used for subsequent analysis. Stiffness of the structure

members at the pile head is also evaluated and appropriate restraint values (translation and rotation) are adopted for detail analysis of the pile.

Design of single pile

Based on design information gathered above, pile design as an isolated pile would be carried out. The aim is to choose suitable pile type, pile section and pile length.

For pile structural analysis as single piles, analytical methods by Hansen & Lundgren (1960) and Viggiani (1981) would have been implemented in the conceptual design stage when assessing general pile stability and adequacy.

For bending moment resistance estimates, an individual pile is first analysed. A program, ALP that analyses a laterally loaded pile with the soil modelled as discrete non-linear springs, and the pile as elastic beam elements, is employed. ALP permits soil displacement to be an input to the program. Other programs with similar capabilities may also be used. The lateral soil pressure response of the pile and structural resistance are important outputs from the analysis. To benchmark the problem, a lateral soil displacement profile is applied to ALP for the selected depth of the slip surface. The quantum of soil displacement is repeatedly increased, until "indicators" in the program show that limiting soil pressures have fully developed along the laterally loaded pile. The pile displacement is then considered as having reached the limiting value parallel to the development of the limiting soil pressure profile at particular locations along the pile. Figures 8.6, 8.7 & 8.8 illustrate the described limiting soil pressures and limiting pile displacements. In the design, either the limiting displacement value is applied, or depending on degree of stability a lower value of displacement (for example values shown in Table 8.1 above) may be applied consistent with the management of the risk of failure.

The next design step is to select suitable pile sections and pile lengths that are capable of withstanding the loads transferred from the structure to the pile head (axial, shear and moment). Checking of the pile shall be carried out for the following conditions:

– At construction stage before the piles are built into the structure. In this case, forces are mainly due to lateral soil movement. As discussed in the Section 8.3, this condition can be critical if the piles are not temporarily tied and braced.
– During service life when the piles are built into the structure. In this case, loading transferred from the structure shall be combined with the loading due to lateral soil movement.

Design of pile group

Next, a check is made on the foundation as a pile group or a frame by applying pile group analysis program such as PIGLET. As a group, the external vertical and horizontal loadings from the superstructure are applied together with the total lateral soil load deduced earlier from the individual pile analysis. The required output is the pile axial loads and the group movement. A frame analysis program with subgrade reaction support (described herein as "Frame" in short) may also be implemented if idealization as a two dimensional case is valid for the group; the advantage of a frame analysis is that the bending moments along the pile shaft may be obtained at the same time. The output as axial response to the lateral soil load and imposed external loads is checked for buckling.

Checking adequate ground support to piles

A check needs to be made of the lateral stability and structural adequacy of the selected pile as an isolated pile to ensure that Mode C of Viggiani (1981) (see Figure 8.9(a) above) prevails in the event of slope movement of the upper layer.

The ground resistance to isolated pile movement during the construction stage (prior to incorporation of the pile head to the substructure and superstructure), is provided by the lower stable layer as the upper layer is moving as a result of slip of the bank. This has been highlighted before when discussing results shown in Figure 8.6. As part of the permanent structure, the resistance to lateral soil load is provided by head restraint for the piles acting as a group, combined with the stiffness of the bottom stable soil layer and the rigidity of the pile.

The vertical bearing capacity of the ground may be estimated by any of the standard procedures. The pile shaft resistance within the moving soil layer has to be ignored. Moreover, settlement consideration can be a governing factor in the pile length design. This will be the case when the soil movement is cyclic and repetitive as would have happened in marginally stable riverbank with high fluctuation of water levels. The soil movement will result repetitive loading and unloading of the piles. Each cycle of such loading will result in a certain amount of residual pile settlement. The cumulated pile settlement after a number of cycles of such loading can be significant. The above implies that, in pile settlement computation, the following settlements should be included:

1. Settlement due to axial load.
2. Cumulated settlement induced in the pile as a result of the cyclic and repetitive soil movement.

The above will normally result in longer pile length required in the lower competent material.

REFERENCES

ALP 1991. *Oasys computer program manual – ALP: Analysis of laterally loaded piles*. London: Oasys Limited.
Broms, B. 1964. Lateral resistance of piles in cohesive soils. *Journal of the Soil Mechanics Division*, ASCE. 90(2): 27–63.
Chan, S.F. 1988. Underpinning of foundation failures using micro piles. *Proceedings of Sarawak Geotechnical Symposium*. Kuching, Sarawak, Malaysia: 59–60.
Chen, L.T. & Poulos, H.G. 1997. Piles subjected to lateral soil movements. *Journal of Geotechnical and Geoenvironmental Engineering*, ASCE. 123(9): 802–811.
Chin Fung Kee, 1979. Course notes on one-day course on piled foundations. The Institution of Engineers Malaysia, Kuching, Sarawak, Malaysia.
Chow, Y.K. & Yong, K.Y. 1996. Analysis of piles subject to lateral soil movements. *Journal of the Institution of Engineers*, Singapore. 36(2): 43–49.
De Beer, E.E. & Wallays, M. 1972. Forces induced in piles by unsymmetrical surcharges on the soil around the piles. *Proceedings of Vth ECSMFE*, Madrid. I: 325–352.
De Beer, E.E. 1977. The state of arts report on piles subjected to static lateral loads. *Proceedings of the Speciality Session 10, IXth ICSMFE*, Tokyo.
Franke, E. 1977. German recommendations on passive piles. *Proceedings of the Speciality Session 10, IXth ICSMFE*, Tokyo.
Hansen, B.J. 1961. "The ultimate resistance of rigid piles against transverse forces". Danish Geotechnical Institute, Bulletin No. 12, Copenhagen: 5–9.
Hansen, B.J. & Lundgren, H. 1960. *Hauptprobleme der bodenmechanik*. Springer-Verlag, Berlin/Gottingen/ Heidelberg: 267–268.
Marche, R. & Scheneeberger, C.E. 1977. Bending moment prediction in piles subjected to horizontal soil movements. *Proceedings of the Speciality Session 10, IXth ICSMFE*, Tokyo.
Ong, D.E.L., Leung, C.F. & Chow, Y.K. 2003. Piles subject to excavation – induced soil movement in clay. *Proceedings of XIIIth European Conference on Soil Mechanics and Geotechnical Engineering*, Prague, Czech Republic. 2: 777–782.
Springman, S.M. & Bolton, M.D. 1990. The effect of surcharge loading adjacent to piles. Transport & Road Research Laboratory, Contractor Report 196.
Stewart, D.P., Jewell, R.J. & Randolph, M.F. 1994. Design of piled bridge abutments on soft clay for loading from lateral soil movements. *Geotechnique*. 44(2): 277–296.
Tan, Y.K. 1988. Abutment failures associated with approach embankments on soft clay. *Proceedings of Sarawak Geotechnical Symposium, Kuching, Sarawak*, Malaysia: 19–31.
Ting, W.H. & Tan, Y.K. 1997. The movement of a wharf structure subject to fluctuation of water level. *Proceedings of XIVth International Conference on Soil Mechanics and Foundation Engineering*, Hamburg.
Ting, W.H. & Tan, Y.K. 2004. Lateral response of piles to soil movement when embedded in banks of limiting stability. Chin Fung Kee Lecture, *Fifteenth Southeast Asian Geotechnical Conference*, Bangkok.
Tschebotarioff, G.P. 1973. *Foundations, retaining and earth structures*. Second Edition, McGraw-Hill Kogakusha Ltd. Tokyo. 396–410.
Viggiani, C. 1981. Ultimate lateral load on piles used to stabilize landslides. *Proceedings of Xth ICSMFE*, Stockholm. 3: 555–560.

CHAPTER 9

Special Topic: Micro Piles

Zamri Chik & Raihan Taha
Department of Civil and Structural Engineering, University Kebangsaan Malaysia

9.1 INTRODUCTION

Micropiles are often used in geotechnical engineering for deep foundation in many parts of the world including countries in the tropics. They are basically smaller diameter piles, usually not greater than 250 mm in diameter, but may occasionally be installed up to 300 mm. (Weltman & Little 1977). Thus, the small diameter feature of the structure naturally accommodates the use of small drilling equipment in soil as well as in rocks. They are mainly used in underpinning works for strengthening foundations. From its early introduction in Europe 55 years ago by Fernando Lizzi, the technique is currently reaching maturity in the other parts of the world including the United States (Bruce 1989) and tropical countries like Malaysia (Chan & Ting 1996). The Root pile, as it was called then, (Bruce et al. 1995) has traversed through a period of innovations to concur with various localized needs.

Other names are also used to imply micropiles, such as Pin-piles, Mini-piles, Pali Radice (Root piles), Fondedile piles, Miga-piles, Pieux Racines, Wurzelpfähle and Estaca Raiz, but all essentially refer to small diameter drilled and grouted piles. The fundamental characteristic lies in its ability to be constructed with equipment used for anchoring and grouting, unlike conventional piles that need to be driven or bored (Bruce & Nicolson 1988).

Besides the introduction of micropiles in difficult ground profiles, the use of micropiles has gained popularity partly because there is a growing trend towards remedial and rebuilding works in older cities and industrial centers. As a result, foundations have to be upgraded or replaced to resist new increased loading, or have to be protected against settlements caused by changes in the soil conditions, for instance as a result of construction of tunnels nearby or other deep excavations.

9.2 GENERAL FEATURES OF MICROPILES

The ability of micropiles in terms of depth is quite considerable. They can be constructed down to depths of 60 meters in some cases through all type of soil, rock and obstructions, and in virtually any angle of installation inclination. They have high slenderness ratio and transfer loads almost entirely by shaft action, eliminating any requirement for under reaming at the base. As all micropiles feature substantial steel reinforcing elements, they can sustain axial loading in both tension and compression. The reinforcement can also be designed to resist bending stresses safely and with minimal displacement. The working load of the piles depends basically on the steel reinforcements and is usually in the range of 10 to 60 tons (89 to 534 kN).

Using micropiles in construction minimizes vibrations, ground disturbance and noise; moreover they can be installed under difficult working conditions. Though in many cases they may be more expensive than conventional driven or large diameter piles, micropiles could be the only guaranteed solution when given a particular set of ground, site access, environmental and performance conditions.

9.3   APPLICATIONS

Main applications include:

– *Bridge piers and abutments* – micropiles may be installed through existing bridge piers and abutments with the minimum vibration, thus achieving a direct, positive connection between pile and structure as illustrated in Figure 9.1(a).
– *Retaining walls* – micropiles systems may be installed to support compressive and tension loads associated with retaining wall instability problems. As the piles are not visible after installation as in Figure 9.1(b), the appearance of the structure is unaltered.
– *Stitch piling* – micropiles are installed through existing walls and foundations to restore or increase the factor of safety. Pile spacing and loads are dependent upon the nature of the existing structure as indicated in Figure 9.1(c).

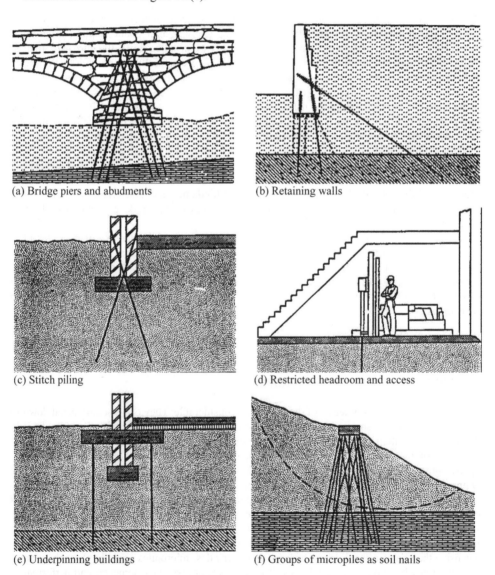

(a) Bridge piers and abudments   (b) Retaining walls

(c) Stitch piling   (d) Restricted headroom and access

(e) Underpinning buildings   (f) Groups of micropiles as soil nails

Figure 9.1.   Some applications of micropiles (Hayward Baker 1986).

– *Restricted headroom and access* – using purpose-built drill rigs, micro piles can be installed in areas of low headroom and very limited excess with the minimum of disturbance as illustrated in Figure 9.1(d).
– *Underpinning buildings* – micropiles used in conjunction with needle beams to support comparatively lightly loaded structures. Pile spacing will be dependent upon the strength of the existing structures. This type of application is demonstrated in Figure 9.1(e).
– Groups of micropiles installed at relatively close centers interact with the surrounding soil to form a composite soil pile retaining structure to stabilize natural and constructed fill slopes. In such cases, micropiles can also be used more directly as "soil nails" to reinforce soil in earth retention and instability in excavations and slopes as illustrated in Figure 9.1(f).
– *Foundations in karstic terrains* – karstic geology is associated with conditions of highly layered rocks with numerous bedding and joint planes as well as the variability of the rock itself. The formations frequently exhibit complex patterns of interconnected cavities of variable size. In such cases, micropiles have been used with satisfactory results from both technical and economical points of view (Uriel et al. 1989).
– Protection of a building during deep excavation in close proximity as shown by the sketches in Figure 9.2(a).
– Protection of buildings in the vicinity of tunneling works as illustrated in Figure 9.2(b).

In another application of micropiles, Flick et al. (1992) described installation of micropiles in Memphis, Tennessee using jet-grouting technique to stop settlement of a building which rested on piled foundations. Moreover, the building was constructed on an ex-landfill consisting of jumbled organic urban waste, trash and soil which include degradable as well as non-degradable materials such as broken concrete slabs, demolition debris, and car bodies.

As micropiles are essentially cast-in-place piles, the soil/pile interaction depends on the mechanical characteristics of the soil, specifically, the outer surface of the pile as well as on the installation procedures. Depending on the type and method of installation, certain type of micropiles can also be classified into displacement micropiles as well as non-displacement micropiles (Lizzi 2000).

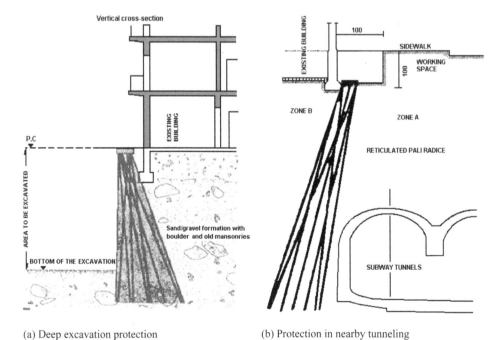

(a) Deep excavation protection   (b) Protection in nearby tunneling

Figure 9.2.   Applications of micropile groups in the protection of structures (Lizzi 1982).

For instance, in the construction of steel shaft micropiles which involves driving of steel shafts into the soil that may cause a greater displacement of the soil. This has constantly been identified as displacement micropiles.

9.4   DESIGN AND CONSTRUCTION CONSIDERATIONS

The design of micropiles follows that of the conventional pile analysis. Therefore soil properties at pile shaft-soil interface and the rock at the tip will have to be carefully characterized. This also includes the evaluation modulus of the soil and the pile (Lizzi 1983). Typical design steps include:

i. geological study;
ii. determination of load to be supported;
iii. design pile-to-structure connection;
iv. design pile-to-soil or rock load transfer; and
v. develop a pile testing program.

However, for patented products, the owner usually suggests various guidelines and certain formulae to be employed in the design.

A typical micropile construction program involves drilling holes by the use of augers, casing, rotary, percussion air or water flushing. Then structural steel piles are lowered as the pile shaft and replacement of reinforcing steel is completed at this stage, if required. Finally, the shaft is filled with cement grout. A set of micropiles is tested up to two times the design load using standard ASTM D1143-81 pile load test criteria. Typical pile test results are shown in Figures 9.3(a) and 9.3(b) (Wrench et al. 1989). Figure 9.3(a) is normally used for the interpretation of the success of the test piles. Figure 9.3(b) is a powerful transformation plot that may be used to evaluate the ultimate capacity of the piles. The plot, also known as the stability plot, originates from Chin and Vail (1973).

9.5   THE DESIGN CONCEPT

As stated earlier the pile can be designed as a friction pile in soils or weathered rocks or as a micropile embedded or plugged in rock formation. Anyway, for the design of micropiles, the end

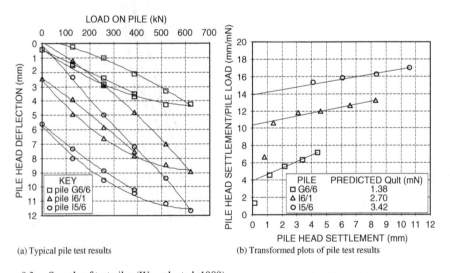

(a) Typical pile test results   (b) Transformed plots of pile test results

Figure 9.3.   Sample of test piles (Wrench et al. 1989).

bearing at the pile tip is not normally considered due to the simple reason that the diameter and hence the area of the base is very small with respect to the pile length. Micropiles can also be designed to carry loads in compression as well as loads in tension.

As there are no design standards specifically for micropiles, standards such as the BS 8110, BS 8081, BS 449 and BS 8004 can generally be referred to. Similarly, guidelines for pile design is available in manuals of practice such as ASCE (American Society of Civil Engineering) Pile Foundations and Pile Structures, Manual of Practice No. 27 and to the more recent ASCE publication in *The Design of Pile Foundations, Technical Engineering and Design Guides* (as adapted from the U.S. Army Corps of Engineers, No. 1). The working stress approach, in view of the compatibility between the structural and the geotechnical designs, is also widely adopted for the micropile design.

Factor of safety (FOS) consideration for both structural and geotechnical capacities is generally taken as two, as the minimum. Such recommendations can be found in the standards for pile design and also in the design of anchors where micropiles are also associated with.

Structural design

Compressive stress in the steel reinforcements used is normally limited in some practices to 50% of the yield strength and the pile capacity is derived from the allowable structural capacity of the reinforcements in the preliminary design. To enhance the allowable structural capacity, other components such as the grout and additional bars may be included. When considering the load transfer at the reinforcement-grout interface, an average ultimate reinforcement-grout bond stress with an appropriate factor of safety is used to derive the required bond length. The following Table 9.1 shows the ultimate bond stresses between cement grout of minimum compressive strength of 30 MPa and steel reinforcement of various types of surface contact and conditions.

The stress distribution along the reinforcement-grout interface is uneven (Woods & Barkhodari 1997). To allow for uncertainties in load distribution along this interface, a minimum bond length of 3 m is suggested. The calculated reinforcement-grout bond length should be checked against the grout-soil/rock bond length. The longer bond length should be taken as the final quantity although the grout-soil/rock bond length could also be optimized by variations in drill-hole size.

If the ultimate stresses given in Table 9.1 are significantly lower than the grout strength, yielding at the reinforcement-grout interface may occur rather than crushing of the grout body when excessive compression is experienced. Assuming that the yield strength of the steel reinforcement is 550 MPa, having the Young's modulus of 210,000 MPa, the strain at 50% yield stress (allowable working stress) will be 1.333E-3. When experiencing similar strain, the grout with the Young's modulus of 28,000 MPa will be under a compressive stress of 37.333 MPa which is greater than the characteristic compressive strength of grout ($f_{cu} = 30$ MPa).

To ensure minimal yielding and that satisfactory load transfer at the pile-soil interface is achieved, the pile axial stress can be kept low to dictate a lower range of strains in the grout. This can be done by increasing the reinforcement or by downgrading the pile load capacity. Similarly, the stiffening effect of the grout in structural confinement either by API pipes, close helical links of the reinforcement bars or by a permanent casing, can substantially reduce the elastic deformation of the micropile under an axial load. Downgrading of the micropile capacity on the other hand or

Table 9.1. Ultimate bond stresses between grout and steel reinforcement (BS 8081).

Ultimate bond stress (MPa)	Contact surface condition
1.0	Clean plain bar or wire
1.5	Clean crimp wire
2.0	Clean deform bar
3.0	Locally nodded strand

increasing the number of piles, in any case may be required in order to comply with a given set of settlement and bearing capacity criteria.

Geotechnical design

Among the factors that influence the ultimate bond stress at the grout-rock are the strength of the grout, rock or soil strength, condition of the contact surface as a result of the drilling, cleaning and grouting operation. In limestone areas in Malaysia, the working bond stress of 0.65 MPa is commonly assumed (Neoh 1996). However, values as high as 1.60 MPa have also been recorded. It is also important to mention that for rock socketed piles, due to a comparatively lower shaft resistance in soil the contribution of the overburden soil to the shaft is usually ignored. Using various methods such as elastic method (Puolos & Davis 1968), finite element method (Osterberg & Gill 1973) and load transfer method (Coyles & Reese 1966) with established database in local soils, the difficulty of assessing the mobilized shaft resistance no longer exists and its contribution can now be estimated more accurately.

For friction micropiles in soil, SPT-N value method (Standard Penetration test) similar to the conventional design method for the cast-*in situ* bored pile can be adopted. It is suggested that this value is equal to twice the SPT-N and not exceeding 120 kPa. Pile capacity and performance of friction piles in soils, weathered and fractured rocks can be improved by pressure grouting. There are mainly two types, namely IRS (Injection, Repetitive and Selection – multi-point and multi-stage grouting) and IGU (Injection, Global and Unitary). IRS is suitable for medium to stiff clayey soil and also well recommended for fractured and karstic formation. It is preferred over IGU due to superior grouting performance as a result of a more stringent procedure. Applied grouting pressure up to 8,000 kPa with water cement ratio of 0.4 to 0.5 have resulted in 50 to 60 percent increase in ultimate bearing capacity of the piles. It is for these reasons that this technique is very popular especially in remedial works involving foundation piles under distress.

9.6   MICROPILE INSTALLATION

Deciding on a particular installation method and execution of proper installation of the piles poses a real challenge to a geotechnical engineer to ensure successful construction. Services of experienced personnel are always an advantage when it comes to determining the rock condition for every pile. The degree of weathering, indicative rock strength, rock mass structures or perhaps karst features may be determined from the excavated rock chipping. Change of hydraulic pressure or sudden drop of drill shaft may indicate karst features, boulders or hard pans. Change in water level or stabilizing fluid may also indicate the presence of a cavity, solution channels and permeable layers where excessive grout loss can be anticipated. The recording of the socket penetration rate is quite important as it could be calibrated to the borehole information and the hydraulic pressure applied on the drill shaft to provide some indication of the rock bed quality.

Drilling method

Although speed of execution is an important criteria in micropile work, the method selected must never compromise the overall structural integrity. Percussion hammer (down-the-hole hammer) or rotary percussive duplex for instance, is recommended in a micropile drilling exercise because of good drillhole protection by advancing steel casings with the hammer simultaneously. Fast penetration, robust, good drillhole verticality, and relatively neat operation (apart from groundwater, no spilling of stabilizing fluid over the work area) are some of its desirable credentials. However, when encountering relatively loose cohesionless soils, there exists a possibility of blowing out of excessive earth materials as it vibrates and advances. It was reported that substantial movements on a bridge structure founded on a sandy stratum was observed when micropiling works nearby employed the percussion hammer method (Shong & Chung 1998). Subsequently minimal movement was recorded as the drilling method was changed to rotary drilling with drilling fluid.

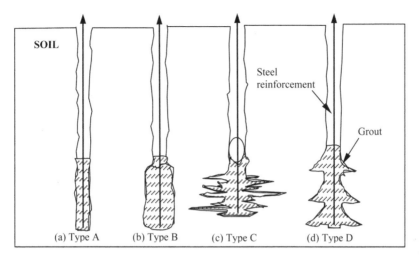

Figure 9.4.   Classification of micropiles based on technique of grouting.

Construction measures

The followings are some suggestions and steps that may result in a better quality of micropile construction.

– Drillhole collapse must be avoided and that can be done by means of a temporary protection casing or/and stabilizing fluid.
– To minimize excessive grout loss, a permanent casing may be used. In cases of sealing a flow channel, the use of rapid hardening grout should be considered.
– Provision of holes should be allowed at the tip of API pipe to facilitate grouting between the drillhole and the pipe.
– It is important to ensure efficient load transfer between the steel reinforcement by having sufficient lapping of the bars. Similarly, proper connection for the coupler and threaded joints is required to guarantee both ends of the pipes are in full contact. Measures should also be taken to avoid weaknesses along these reinforcements.
– Although cleaning at the inner surface of the pipes is very inconvenient, to ensure good bonding, grease or other coating on the reinforcement should be removed.
– Excessive heat can alter the physical and chemical properties of the high yield reinforcement bars, hence precaution should be taken when welding works are required.
– Proper placement of the centralisers of micropile reinforcement is important to ensure adequate grout cover for the bonding of the interfaces. And it is also advised that grouting be carried out immediately after cleaning/flushing of the drillhole with clean water.

The grouting technique is probably the most sensitive construction control over soil-grout interface bond capacity as it directly affects the outcome of the bonding. Micropiles have since been classified into Type A, Type B, Type C and Type D as indicated in Figure 9.4 based primarily on the method of placement and pressure under which grouting is used during the construction. Other commonly known micropiles such as Menard and Gewi piles are also illustrated in Figure 9.5. Figure 9.6 shows the construction of a micropile system using API pipe compared to a reinforcement bar system.

The classification as shown in Figure 9.4 is described as follows:

Type A: The grout column is not pressurized. Sand-cement mortars as well as neat cement grout is placed under gravity head only.

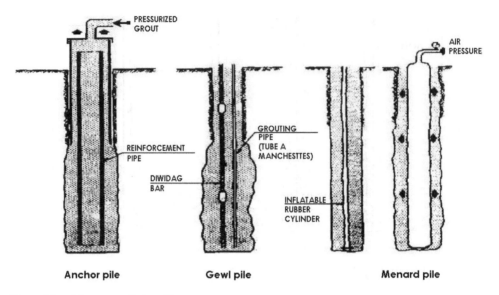

Anchor pile   **Gewl pile**   **Menard pile**

Figure 9.5. Other types of micropiling.

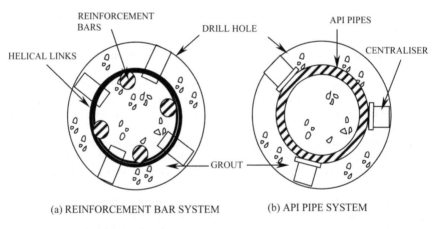

(a) REINFORCEMENT BAR SYSTEM   (b) API PIPE SYSTEM

Figure 9.6. Cross-section of two common micropile systems.

Type B: As the temporary steel drill-hole casing is withdrawn, the neat cement grout is placed under pressure in the range 0.5 to 1.0 MPa to maintain a seal around the casing during its withdrawal and is limited to avoid excessive grout takes.

Type C: The grouting involves 2 stages. The primary grout is placed under 1.0 to 2.0 MPa grouting pressure causing hydrofracturing of the surrounding ground. Next, prior to hardening of the primary grout (a lapse of 15 to 25 minutes), grouting is again carried out (called secondary grout) via "tube-a-machette". Sometimes this technique is referred to as IGU (Injection Globale et Unitaire).

Type D: This is similar to the 2-step grouting as in Type C with some modification. Hardening of the primary grout is allowed before the secondary grouting is performed. The additional grout is injected via "tube-a-machette" under high pressure of 2.0 to 8.0 MPa. A packer may be used so that specific levels can be treated several times, if necessary.

Table 9.2.   Two micropiles test results.

| Test pile | Settlement of the pile head (mm.) | | | |
	1 × W.L. (= 12000 kN)	First unloading	2 × W.L. (= 24000 kN)	Second unloading
TP-B22	6.7	2.6	20.4	4.5
TP-C27	6.3	0.3	20.1	1.5

9.7   SOME CASE HISTORIES

Case history I

In this case 2 micropiles subjected to a working load of 1200 kN were each tested in a different founding formation. They were 250 mm in diameter and were reinforced by Grade N80 API pipe of 177.8 mm diameter and 10.36 mm thickness. A 267.4 mm diameter permanent casing with 16.6 mm thickness was also installed for both structures extending to the completely weathered granite for the Test pile B22 as a soil friction pile and the other Test pile C27 was rock socketed in solid limestone to a depth of 8 m. The total length of Test pile B22 was 54 m and Test pile C27 was 35 m. As indicated in Table 9.2, the test results are almost identical although the friction pile TP-B22 exhibits slightly higher settlement readings after the loading and unloading. It was also observed that there was almost no residual settlement during the first cycle of unloading.

Case history II

An old property in Boston needed additional support due to increased loading from redevelopment and refurbishing of its six-storey structure (Bruce & Nicholson 1988). The fill consisted of saturated loose-gray brown fine sand and silt, and overlaid soft gray organic silt with traces of shells, sand and gravel.

Piling had to be installed inside a partially demolished basement with only 8 feet of headroom. The difficult and restricted access ruled out the use of conventional piles. Based on cost and performance analysis, about 260 mini-piles with working loads of 40 tons in compression and 12 tons in tension were installed. Prices were around USD65 per foot run (in 2003) and this was much more than the conventional piling system. However, the piles provided more support per foot, so fewer feet were needed.

The founding layer occurred at about – 4 ft. and was 18 to 24 ft. throughout the site. This consisted of medium dense/ dense fine-medium sand with a trace of silt. The pile lengths were kept within this horizon so as not to penetrate the weaker underlying Boston blue clay.

The features of the installed micropiles were as follows:

– the design was based on ultimate load which is 2.3 times the design working load: 92 tons (819 kN) in compression and 27 tons (240 kN) in tension.
– ultimate load of 92 tons required a load transfer length of 15.5 feet.
– a diesel hydraulic track-rig installed the piles.
– steel casing was 5.5 in. diameter with 0.362 in. wall thickness with a minimum specified yield stress of 55 ksi.
– the rebar was 1 in. diameter and 60 ksi yield stress.
– type 1 grout with w/c ratio of about 0.5 were placed by tremie.
– grouting pressure was 60 ksi (414 MPa).
– 14-day strength of concrete cube was over 6000 psi (41.4 MPa).
– 2 piles were test loaded to compression and tension. The elastic settlement at 80 tons was about half the total deflection. No evidence of pile or soil failure. Net butt settlement was well below recommended level. Performance in tension was equally satisfactory.

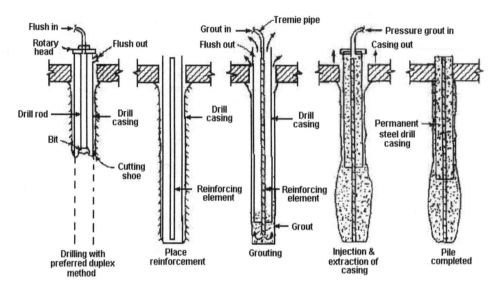

Figure 9.7.   Construction stages of Boston pin pile (Bruce & Nicholson 1988).

Figure 9.7 shows the construction stages of the micropile installation for this project. The installation took 3 months and the major structural rebuilding took 8 months. The performance of the piles was very satisfactory.

Case history III

In this case, micropiles were used to support part of the I-78 dual highway, which crosses the Delaware River between Pennsylvania and New Jersey. It was thought that the New Jersey pier was to be founded on solid rock but excavation revealed that the solid rock was nonexistent. So two alternatives were considered: the installation of six 36 in. caissons, each with a working load of 360 tons (3,203 kN); and the placement of 24 minipiles, each of nominal working load of 100 tons (890 kN).

Because of the geology of the area, the installation of the caissons might delay the overall construction time dramatically. In addition, the costs of the caissons were difficult to estimate due to wide variations in the design and construction as different soil and rock conditions were encountered.

On the other hand, the pin pile cost was in excess of USD100 per foot. Though expensive, they did offer a number of advantages that made their use attractive and cost effective. Because pin piles transfer their loads by skin friction, there was no possibility that the pile could fail by punching through into a soft underbed immediately under the founding level. Even the diverse soil conditions found in this site could not dramatically alter construction times. In addition, test piles could prove their effectiveness in advance.

The geology of the area consists of bedrock from Cambrio-Ordovician dolomite limestone. It proved to be moderately to high fissured, and major clay filled beds intersected the bedrock even at depths greater than 100 ft.

The features of the micropiles were:

- the design pile diameter was 8.5 in. and 15 ft. length in competent rock which was based on 100 tons working load.
- the major structural component of the piles was a 55 ksi (379 kPa) low alloy steel pipe with 7 in. outside diameter and a 0.408 in thick wall.
- a separate test pile, 30 ft. long with 5.33 ft. of bond was loaded to 205 tons. This pile was chosen because it met the owner's design requirement for rock grout bond of 50 psi. Total settlements

of 0.367 in. and 0.373 in. were recorded at each successive cycle to 205 tons. Creep at 0.011 in. was recorded over 1 hour at these loads. The permanent set after this operation was 0.07 inch. When tested to higher level, the upper casing began to buckle at 224 tons (1993 kN).

– grouting pressure was 50 ksi (345 MPa).
– following the test, it was decided that seven of the piles were required to have continuous 15 foot bond zones. But because the rock was likely to be variable, the 15 feet bond zone did not have to be continuous for the rest of the piles. Nevertheless, they would have to meet other requirements, i.e., the lower part of the zone had to contain at least 10 feet of continuous sound rock. A zone of acceptable load bearing rock had to be at least 5 ft thick.
– overall, a total of 1710 linear feet of piles was installed, with length of individual piles ranging from 30 ft to 43 ft more than projections. A volume of grout equivalent to 8 times the nominal drilled-hole volume was injected. Much of this was consumed in the zone above the rock head during grouting operation.

The piles performed extremely well after installation.

9.8   CONCLUSIONS

Micropiles are smaller diameter piles mainly used in an underpinning operation when there are no other options to support buildings in use. A wide range of applications has also been documented which shows the versatility of this technique. The piles can therefore function like an ordinary foundation pile as well as a compensation pile for remedial works especially in areas with site constraints. The design procedures for micropiles are no different than the conventional pile although various products have their own guidelines.

Care must be taken during the construction against hole collapse because such events will lead to the loss of strength of the grout. Borehole debris that is left behind can lead not only to the loss of end bearing but also a significant part of shaft friction in the bearing stratum. To avoid failure, special attention needs to be given to providing high quality grouting. There is evidence that the strength of the grout is highly enhanced by pressure grouting techniques. Even though micropiling work is comparatively expensive, it, offers a number of advantages that make its use attractive and also cost effective.

REFERENCES

American Society of Civil Engineers. 1959. Pile foundations and pile structures, Manual of Practice. (27): 72.
American Society of Civil Engineers. 1993. Design of pile foundations (Technical Engineering and Design Guides as Adapted from the U.S. Army Corps of Engineers, No. 1), *American Society of Civil Engineers*, New York.
American Society for Testing and Materials (ASTM) Specification. 1981. Piles under axial compressive load, *ASTM D 1143*, Part 19, Philadelphia, PA.
British Standards Institution. 1970. British standard code of practice for the specification for the use of structural steel in building, *BS 449*: 1970.
British Standards Institution. 1985. British standard code of practice for the structural use of concrete (formerly CP 110), *BS 8110*: 1985.
British Standards Institution. 1986. British standard code of practice for foundations (formerly CP 2004), *BS 8004*: 1986.
British Standards Institution. 1989. British standard code of practice for ground anchorages, *BS 8081*: 1989.
Bruce, D.A. & Nicholson, P.J. 1988. Minipiles mature in America, Civil Engineering, *ASCE*. 58(12): 57–59.
Bruce, D.A. 1989. American development in the use of small diameter inserts as piles and in situ reinforcement, *Proceedings of International Conference on Piling and Foundation*. 1: 11–22.
Bruce, D.A., DiMilio, A.F. & Juran, I. 1995. A primer on micropiles. Civil Engineering, *ASCE*, December: 51–54.
Chan, S.F. & Ting, W.H. 1996. Micropiles. *Proc. of the 12th Southeast Asian Geotechnical Conference*, Kuala Lumpur. 2: 93–99.

Chin, F.K. & Vail, A.J. 1973. Behavior of piles in alluvium. *Proceedings 8th ICSMFE*, 2.1: 47–52.

Coyles, H.M. & Reese, L.C. 1966. Load transfer for axially loaded piles in clay. *Journal of Soil Mechanics and Foundation Division, ASCE*. 92(SM2): 1–26.

Flick, L.D., Osborn, N.B.R., Graham, A.E., Marasa, M.J. & Tobey III, F.T. 1992. Minipiles Limestone, Civil Engineering, *ASCE*, Sept: 46–49.

Hayward Baker, 1986. *Minipiles in ground modification*. GKN Hayward Baker Inc.

Lizzi, F. 1982. *The static restoration of monuments*. Genoa: Sagep Publisher.

Lizzi, F. 1983. The raticolo di pali radicè (Reticulated root piles) for the improvement of soil resistance. Physical aspects and design approaches, *Proceedings 8th ECSMFE*, Helsinki. Rotterdam: Balkema.

Lizzi, F. 2000. Micropiles: Past; present ... and future, *Proceedings of the GEOTECH – YEAR 2000: Developments in Geotechnical Engineering*, Bangkok, Thailand: 145–152.

Neoh, C.A. 1996. Performance of slender micropiles socketed in limestone bedrock, *12th Southeast Asean geotechnical conference*, Kuala Lumpur, Malaysia: 493–501.

Osterberg, J.O. & Gill, S.A. 1973. Load transfer mechanism for piers socketed in hard soils or rocks, *Proceedings of the 9th Canadian Rock Mechanics Symposium*, Montreal: 235–261.

Poulos, H.G. & Davis, E.H. 1968. The settlement behavior of singly axially loaded incompressible piles and piers, *Geotechnique*. 18: 351–371.

Shong, L.S. & Chung, F.C. 1998. Design and construction of micropiles. *Proceedings for A two-days Conference on Geotechnical Engineering, IEM*, Kuala Lumpur: 201–223.

Uriel, A.O., Ortuno, L. & Puebla, F.J. 1989. Micropiles for building foundations on karstic areas. *Proc. 12th ICSMFE*, Rio de Janeiro, Brazil. 2: 1039–1042.

Woods, R.I. & Barkordari, K. 1997. Load transfer mechanism in ground anchorage, *Numerical modes in geomechanics*, Balkema.

Weltman, A.J. & Little, J.A. 1977. A review of bearing pile types, DOE and CIRIA piling development group report PG1, London: 82.

Wrench, B.P., Heinz, H. & Salerno, G. 1989. "Underpinning a multi-story building using micropiles", *Proc. 12th ICSMFE*, Rio de Janeiro, Brazil. 2: 1043–1047.

CHAPTER 10

Special Topic: Innovative Pile Foundations

Mageswaran Pavadai
R & A Geotechnics Sdn. Bhd., Kuala Lumpur, Malaysia

Bujang B.K. Huat
Department of Civil Engineering, University Putra Malaysia, Malaysia

10.1   INTRODUCTION

In recent years, there have been a number of innovative piling systems in addition to the already established driven piles, drilled shafts and caisson foundations. A variant of the pile support system is the Augeo pile system. According to Abdullah et al. (2003), the Augeo pile system is theoretically cheaper than the conventional pile system.

10.2   THE AUGEO PILE SYSTEM

The Augeo piled embankment system consists of lightweight piles with an enlarged individual pile cap and pile-foot or pile base. A gravel mattress wrapped in two layers of high strength geogrids is placed on top of the pile caps to transfer the load of the embankment to the pile caps (Figure 10.1). The piles are pushed into the soil in a close square grid using modified vertical drain stitcher equipment at a high speed. In this manner, the loads are theoretically directly transferred to the competent layers (with SPT $N > 12$), and the soft compressible layers are not loaded. The excessive settlements are avoided and construction time can be limited.

The pile system consists of cement mortar filled plastic casing as piles (Figure 10.2), sand blanket, pile cap, and granular mattress with geogrids layers.

A typical 2-meter high embankment built on the Augeo pile system is schematically presented in Figure 10.3.

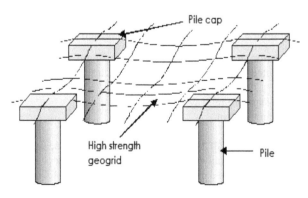

Figure 10.1.   Schematic view of the Augeo pile system.

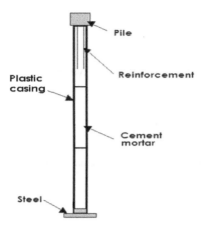

Figure 10.2.   The Augeo pile.

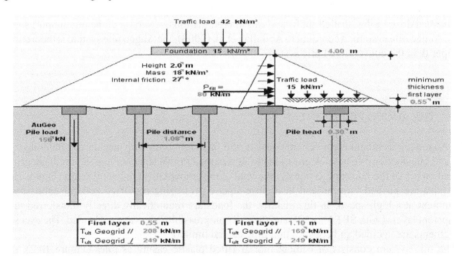

Figure 10.3.   Typical section of the Augeo pile system (from Abdullah et al. 2003).

10.3   DESIGN OF THE AUGEO PILE

The design of a typical railway track embankment supported on Augeo piles is shown in an illustration. Figure 10.4 shows the typical embankment section.

The design calculation for a railway must follow strict design criteria that are listed as follows:

– Settlement free foundation (max allowable settlement <25 mm).
– Settlement free compacted fill for embankment construction.
– Life time > 100 years.
– 2 tracks for electrified standard railway traffic.
– Width top of embankment ±15 m.
– Embankment side slopes 1:2.
– Train load according to European Standard EN 1991-2 Eurocode 1, Section 6, Rail traffic actions and other actions specifically for railway bridges and approaches. Section 6 of Eurocode EN1991-2 describes a 4-point load each 250 kN center-to-center; 1.6 m. 0.8 m from each point load to total live load of 80 kN/m per track. Effective width is 3 m on 0.7 m depth (base ballast bed). This equals a maximum traffic load of 52 kN/m². Due to spreading of the load in the

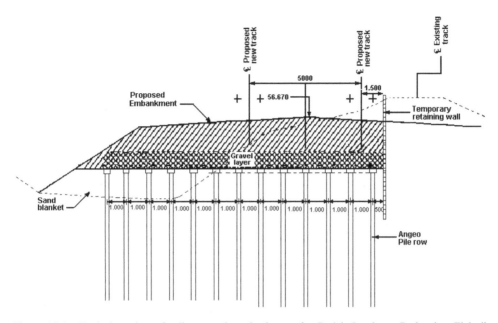

Figure 10.4.   Typical section of railway track embankment for Projek Landasan Berkembar Elektrik Rawang – Ipoh (Client – Keretapi Tanah Melayu) (after Cortlever 2001).

embankment, this value will be reduced by 4 kN/m² over the first meter of the embankment height, and furthermore, decreases by 3 kN/m² per 0.5 m increase of the embankment height.

– Load of ballast bed = 15 kN/m² based on a width of 5 m.
– According to the BS8006, the following load factors should be applied:
  Embankment fill: $f_{fs} = 1.3$
  Dead loads: $f_f = 1.2$
  Live loads: $f_q = 1.3$
– Construction time of 10 months.
– Rest settlement over 24 months <25 mm.
– Design standard BS8006 for embankment, piles and geogrid.
– Weight of fill material – 18 kN/m³.
– Modulus of elasticity of fill material >18,000 kN/m².

In addition to the above, the design of the Augeo pile embankment should follow the guidelines as stipulated in BS8006 with the following listed parameters taken into account:

– Max. allowable pile load   150 kN
– Mass of the fill material   18 kN/m²
– Internal friction fill material   27°
– Traffic load construction layer   15 kN/m²
– Weight ballast bed   15 kN/m²
– Pile cap size   300 mm × 300 mm

Based on the above design parameters and design criterion, we shall proceed with embankment design calculation which consists of the following major steps.

Step 1: Computation of Augeo pile spacing

The pile spacing, s, is computed based on the formula below:

$$s = \sqrt{\frac{F_p}{(f_{fs} \times H \times \gamma) + (f_f \times w_s) + (f_q \times q_s)}} \qquad (10.1)$$

Table 10.1. Pile spacing for varying embankment heights.

Embankment height (m)	Traffic load distribution at cap level (kN/m^2)	Pile distance, s in square pattern based on load schedule (m)
1.0	48	1.15
1.5	45	1.12
2.0	42	1.08
2.5	39	1.05
3.0	36	1.03
3.5	33	1.00
4.0	30	0.98

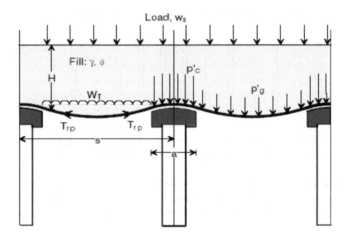

Figure 10.5. Load scheme (from Figure 70 Section 8 BS8006: 1995).

where F_p = pile working load (kN); H = embankment height (m); γ = unit weight of fill (kN/m^3); w_s = traffic load (kN/m^2); and q_s = dead load (kN/m^2).

Load factors as per BS8006 Table 27 for Ultimate Limit State Conditions are:

Load factor of embankment fill: $f_{fs} = 1.3$
Load factor of dead loads: $f_f = 1.2$
Load factor of live loads: $f_q = 1.3$

Based on this formula, the pile spacing, s, in square pattern is calculated for the center of the embankment with heights varying from 1 to 4 m in steps of 0.5 m. The results of the analysis are tabulated in Table 10.1 above.

From Table 10.1, the pile spacing can be increased for the piles below the side slope. For economic construction purposes, the piles which fall below the side slope are slightly increased based on the lateral pile extent and final embankment height.

Step 2: Vertical load shedding

This step is undertaken to evaluate the vertical stress on the pile cap and interconnecting base geogrid. Arching rate based on BS8006 is slightly modified to suite rectangular pile spacing scheme. The vertical load shedding scheme is described in Figure 10.5.

Average vertical stress in soil, σ'_v:

$$\sigma'_v = (f_{fs} \times H \times \gamma) + (f_f \times w_s) + (f_q \times q_s) \tag{10.2}$$

Average vertical stress in soil, p'_c (Sprangler & Handy 1973; John 1987):

$$p'_c = (C_c \times \frac{a}{H})^2 \times \sigma'_v \tag{10.3}$$

where C_c = arching coefficient for concrete piles = $1.95\frac{H}{a} - 0.18$; and a = width of pile cap (mm).
Vertical load on interconnecting geotextile/between the pile cap, $W_T(kN/m)$:

$$W_T = \left(\frac{s(f_{fs}\gamma H + f_q w_s)}{(s^2 - a^2)} \right) \times \left(s^2 - a^2 \left(\frac{p'_c}{\sigma'_v} \right) \right) \quad \text{when } 0.7(s - a) < H < 1.4(s - a)$$

$$W_T = \left(\frac{1.4 s f_{fs}\gamma(s - a)}{(s^2 - a^2)} \right) \times \left(s^2 - a^2 \left(\frac{p'_c}{\sigma'_v} \right) \right) \quad \text{when } H > 1.4(s - a)$$

$$W_T = 0 \quad \text{when } (s^2 - a^2) < \frac{p'_c}{\sigma'_v}$$

Step 3: Design of geogrid

Partial Safety Factor as recommended by BS8006 for usage of geogrid:

Partial factor manufacturing	f_{m10}
Partial factor extrapolation test data	f_{m12}
Partial factor installation damage	f_{m21}
Partial factor degradation	f_{m22}
Partial factor safety	f_m
Reduction factor for creep	f_{cr}

Average tensile load on geogrid, T_{rp} (Leonard 1988),

$$T_{rp} = \frac{W_T(s - a)}{2a} \times \sqrt{\left(1 + \frac{1}{6\varepsilon} \right)} \tag{10.4}$$

where ε = strain in geogrid (%) provided by the supplier ($\varepsilon = 5.5\%$ for ultimate limit state).
Required breaking strength of geogrid, T_D

$$T_D = T_{rp} \times f_m \tag{10.5}$$

where $f_m = f_{m10} \times f_{m12} \times f_{m21} \times f_{m22}$
Ultimate breaking strength of geogrid, T_{ult}

$$T_{ult} = T_D \times f_m \times f_{cr} \tag{10.6}$$

The strength of geotextile is finally chosen based on ultimate breaking strength, T_{ult} derived from earlier mentioned limit state condition where the most unfavorable value is taken. Table 10.2 below is derived from the basis of the above mentioned calculation showing corresponding pile spacing along and perpendicular to the track and the required geogrid reinforcement with varying embankment height.

Based on these results, the geogrid strength is chosen combined with 300 mm pile caps under the slopes as well as the center of the embankment. The pile rows in the direction of the embankment vary depending on the embankment height.

Table 10.2.   Pile spacing and geogrid for varying embankment heights.

	Pile spacing					Required geogrid (kN/m)	
	Rectangle pattern						
			Perpendicular track			Perpendicular track	Along track
Height fill (m)	Square pattern (m)	Along track (m)	Center (m)	Slopes (m)	Height fill (m)	Geogrid	Geogrid
6.0	0.90	0.80	1.00	1.25	6.0	400	100
5.5	0.92	0.84	1.00	1.25	5.5	350	100
5.0	0.94	0.88	1.00	1.25	5.0	300	100
4.5	0.96	0.92	1.00	1.25	4.5	250	150
4.0	0.98	0.96	1.00	1.25	4.0	250	150
3.5	1.00	1.00	1.00	1.25	3.5	250	200
3.0	1.03	1.05	1.00	1.25	3.0	250	200
2.5	1.05	1.10	1.00	1.25	2.5	250	250
2.0	1.08	1.17	1.00	1.25	2.0	250	300
1.5	1.12	1.24	1.00	1.25	1.5	250	350
1.0	1.15	1.32	1.00	1.25	1.0	250	400

This geogrid is located on the load spreading layer consisting of a 600 mm thick gravel layer. The function of the mattress is three-fold:

- Create a mattress that can transfer loads from the embankment to the piles.
- Take care of the anchor loads at the border of the embankment.
- Improves the shear strength between geogrid and soil.

The lower geogrid layers run in two directions to provide adequate vertical load shedding to the pile caps. The upper geogrid layer is a composite of a geogrid and a geotextile. The geotextile is to prevent migration of fines in the gravel bed. The design of the pile pattern and the geogrids are based on the following assumptions:

- The pile caps are in line.
- The required force in the geogrids will not fluctuate too much over the width of the embankment.

Step 4: Pile group extent

The piled area should extend to a distance beyond the edge of the shoulder of the embankment to ensure that differential settlement or instability outside the piled area will not affect the embankment crest. The edge limit of the outer pile cap is given by the following equation (see also Figure 10.6) (Carlsson 1987).

$$L_p = H(n - \tan \theta_p) \quad \text{and} \quad \theta_p = 45° - \frac{\varphi_{cv}}{2}$$

Based on these equations the minimum distance between the edge of the shoulder and the outer limit of the pile cap is given in following Table 10.3.

Step 5: Reinforcement bond

The reinforcement should achieve an adequate bond with the adjacent soil at the extremities of the piled area. This is to ensure that the maximum limit state tensile loads can be generated (across

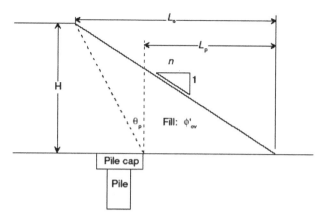

Figure 10.6.   Notation for outer limit of pile caps (from Figure 69 Section 8 BS8006: 1995).

Table 10.3.   Values of L_p for corresponding
embankment heights.

Embankment height at outer edge (m)	Distance L_p (m)
1.50	2.08
1.75	2.43
2.00	2.77
2.25	3.12
2.50	3.47
2.75	3.81
3.00	4.16
3.25	4.51
3.50	4.86
3.75	5.20
4.00	5.55

the width and along the length of the embankment) between the outer two rows of piles. The bond length can be calculated with the following equation:

$$L_b = \frac{f_n f_p T_{rp}}{\gamma h \dfrac{(\alpha'_1 \tan \varphi'_{cv1})}{f_{ms}} - \dfrac{(\alpha'_2 \tan \varphi'_{cv2})}{f_{ms}}} \tag{10.7}$$

where T_{rp} = load in geogrid due to vertical load shedding; L_b = bond length beyond outer pile row; f_n = partial load factor economic ramification; f_p = partial load factor pull-out resistance; h = average height of fill over the bond length; γ = unit weight of the embankment; α'_1 = interaction coefficient fill; α'_2 = interaction coefficient gravel; f_{ms} = partial material factor applied to tan φ_{cv}; φ_{cv} = internal friction of fill.

Based on this equation the following bond lengths are calculated:

Surcharge Load (kN/m^2)	// Pile distance (m)	/ (m)	Height h (m)	// T_{rp} (kN)	L_b (m)	/ T_{rp} (kN)	L_b (m)
63.0	1.30	1.00	1.00	139	5.51	102	4.05
61.5	1.25	1.00	1.25	131	5.04	100	4.23
60.0	1.25	1.00	1.50	131	4.90	122	4.56
58.5	1.20	1.00	1.75	102	3.71	133	4.83
57.0	1.15	1.00	2.00	102	3.61	142	5.02
55.5	1.15	1.00	2.25	102	3.51	154	5.30

Note: // – parallel to the track, / – perpendicular to the track.

The BS8006 is use to determine the forces in the geogrids in the piled embankment. However to calculate the loads on the pile caps, a finite element calculation program, such as PLAXIS has to be used to determine loading spreading in the embankment. Then it will be possible to create variations in the pile distance in the cross direction as well. The pile distance and geogrid forces are dictated by the maximum allowable pile load.

10.4   CONSTRUCTION OF AUGEO PILE EMBANKMENT

The construction of a typical a Augeo pile embankment is illustrated as follows:

(a) Corrugated HDPE pipe is capped with 230 × 230 mm steel plate at bearing end and pushed into the soft ground with the aid of a 220 mm radius mandrel.

(b) The fabricated Corrugated HDPE casing is slid into the modified stitcher machine to push the pipe with the aid of a mandrel into the ground, up to two times the pile computed working load. The steel mandrel is then retracted leaving behind the corrugated pipe.

Figure 10.7.   Construction of a typical Augeo pile embankment (www.cofra.com).

(c) Upon installation to the required pile spacing, the depth of pile is measured and reinforcement is lowered into the casing to the soffit of the pile.

(d) The HDPE corrugated pipe is filled with grade 30 mortar and allowed 7 days for curing.

(e) The cured pile is capped with precast $300 \times 300 \times 150$ mm thick grade 30 concrete cap.

(f) Upon completion of the cap installation, sand is leveled between the caps to the top of cap level before placement of geogrid.

(g) Geogrid layer is placed over the sand bedding acting as the mediator to spread the embankment pressure/load onto the piles.

(h) Gravel bed is laid between two layers of geogrid before proceeding with layering suitable earth fill to form the final embankment profile.

Figure 10.7.   (Continued)

(i) Complete Augeo pile embankment safe for dedicated use.

Figure 10.7. (Continued)

REFERENCES

Abdullah, A., John, A.N. & Arulrajah, A. 2003. Augeo pile system used as piled embankment foundation in soft soil environment. *Proceedings of 2nd International Conference on Advances in Soft Soil Engineering and Technology*, Putrajaya, Malaysia: 703–714.

British Standards BS 8006: 1995. Section 8. Design of embankments with reinforced soil foundations on poor ground. *British Standard Institution*, UK.

BS EN 1991-2 Eurocode 1, Section 6. Actions on structures. Traffic loads on bridges. *British Standard Institution*, UK.

Carlsson, B. 1987. *Armerand Jord*.Terranova, Sweden.

Cortlever, N.G. 2001. Design of double railway track on Augeo Pile System. *Symposium on Soft Ground Improvement and Geosynthetic Application*, AIT, Bangkok.

John, N.W.M. 1987. *Geotextiles*. Glasgow: Blakie.

Leonard, J.W.1988. *Tension structures*. New York: McGraw Hill.

Sprangler, M.G & Handy, R.L. 1973. *Soil engineering*. New York: Intext Education Publishers.

CHAPTER 11

Special Topic: Piles Load and Non Destructive Test

Bujang B.K. Huat, Shukri Maail & Azlan A. Aziz
University Putra Malaysia, Malaysia

11.1   INTRODUCTION

After the piles have been installed inside the ground, a number of these piles would be selected for testing. Pile needs to be tested, as its capacity could not be ascertained just by design and construction processes only. The condition and strength of soil need also to be taken into account. In difficult sites, the possibility of piles being damaged or having defects can only be determined from the load test.

The main objectives of performing the test can be listed as follows:

1. To prove that the pile will not fail at working load.
2. To determine ultimate bearing capacity of the soil. This can also be used to check calculation of the pile design.
3. To obtain a relation between load and settlement, especially at working load.
4. To check the quality of work.

Two pile testing methods are normally carried out, namely.

1. The full scale static load test on prototype foundations, and
2. Dynamic methods, which are based on the dynamics of the pile driving or wave propagation.

11.2   STATIC LOAD TEST

A precise way to determine the ultimate axial (as well as pull out) of deep foundations is to build a full-size prototype foundation at the site of the proposed foundation and slowly load it to failure. This method is known as static load test. Static load tests are generally much more expensive and time-consuming, and thus must be used judiciously.

The objective of a static load test is to develop a load-settlement curve or, in the case of uplift tests, a load-heave curve. This curve is then used to determine the ultimate load capacity.

Two test methods are normally used, that is the maintain load test and constant rate penetration test. Each of these tests have their own merits and demerits.

Maintain Load Test.   In this test, load is applied in stages according to specific time periods. When the load is maintained, the value of pile settlement is observed and recorded. This settlement observation is made at all stages of the applied load. This load test is suitable for determining the acceptable value of piles and to obtain a relationship between pile settlement and load. As this test may require holding each load for at least 1 or 2 hours, sometimes more, the tests may require 24 hours or more to complete.

Constant Rate Penetration Test.   In this test, the pile is loaded in such the way that the rate of penetration of the pile is constant. A reasonable constant rate is between 0.4 to 2.0 mm/min. While the pile is inserted at a constant rate, the load is observed. Typically each load increment is about 10% of the anticipated design load and held for 2.5 to 15 minutes. The process continues until

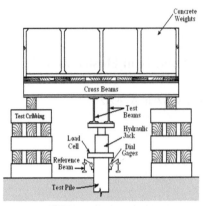

(a) Schematic diagram showing the system.

(b) Photograph of the kentledge system.

(c) Load cell.

Figure 11.1.   Use of a hydraulic jack reacting against dead weight to develop the test load in a static load test.

reaching about 200% of the anticipated design load of "failure" and generally requires 2 to 5 hours to complete. Compared with the maintain load test, this method is faster and suitable for determining the ultimate load of a pile.

To conduct a static load test, there must be a means of applying the desired loads to the foundation and measuring the resulting settlement. The most common method is the kentlegde system whereby dead weights such as precast concrete blocks are stacked on top of the foundation, as shown in Figure 11.1. An alternative is to provide multiple support in the form of reaction pile and use them as a reaction for a hydraulic jack (Figure 11.2).

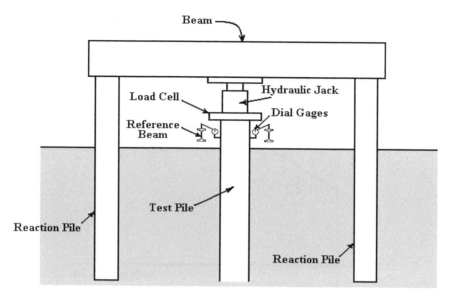

Figure 11.2.   Use of a hydraulic jack reacting against a beam and reaction piles to develop the test load in a static load test.

Interpretation of static load test results

From the load-settlement curve obtained from the static load test, the ultimate load capacity of the pile can be determined. This, however, requires where the "failure" occurs to be defined, i.e. for cases that do not exhibit a clear plunging failure. Various methods are available to estimate the ultimate pile capacity such as Chin's (1970) method and Davisson's (1973) method.

Chin's method

This method was introduced by Chin Fung Kee in 1970, and had been widely used in Malaysia. The method assumes that the shape of the load-settlement curve is in the shape of a hyperbola that can be expressed as

$$\Delta/Q = m\Delta + c \tag{11.1}$$

where, Q = pile load; Δ = pile settlement and m and c are constant.
   If the above equation is divided with Δ, it becomes

$$1/Q = m + c/\Delta \tag{11.2}$$

From Equation 11.2, it is found that $1/Q$ is equal to gradient m. Q can be therefore be assumed to equal $1/m$ if the term c/Δ is ignored.
   Figure 11.3 shows a typical pile load settlement graph plotted in terms of Δ/Q versus Δ. As shown, two distinct straight lines were obtained. The first line has a gradient of m_1 representing pile at the early stage of the load test. As this stage, it can be assumed that only the pile skin friction (Q_s) is mobilized, i.e.

$$Q_s = 1/m_1 \tag{11.3}$$

while the second line has a gradient m_2, representing the overall pile capacity, i.e. with both the skin friction and end bearing fully mobilized. The ultimate capacity of the pile (Q_t) is therefore

$$Q_t = 1/m_2 \tag{11.4}$$

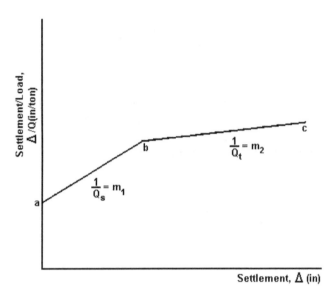

Figure 11.3.   Chin's method (after Chin 1970).

Davisson's method

Davisson's (1973) method is one of the most popular methods in the US (Coduto 2001). It defines the pile ultimate capacity as that which occurs at a settlement of $4\,\text{mm}\,(0.14\,\text{in}) + d/120 + QL/(AE)$ as shown in Figure 11.4. The Davisson's method seems to work best with data from the maintain load test.

Pile settlement acceptance criteria

Apart from satisfying the required factor of safety for pile capacity, acceptance of a working pile is also based on the specified settlement performance of the pile. These acceptance criteria may vary depending on the requirements of the specific project. In general, the satisfaction of these acceptance criteria would indicate an acceptable and adequate pile design, adopted method of pile installation and quality of construction.

There are at present various sets of pile acceptance criteria in use depending on local and/or regional preferences by the engineering community. These settlement acceptance criteria are specifically for single vertical piles under static axial loading conditions. In general, these settlement acceptance criteria specify the acceptable limiting pile settlement for the following conditions, namely:

a. At the specified service working load level,
b. At two times the service working load level, and
c. The residual pile settlement upon unloading to zero load level.

The residual settlement criterion is an important aspect that has a significant impact on the performance acceptance of a single pile test. Two conditions of assessment of the pile residual settlement are considered at present, namely:

a. Upon unloading from the service working load level and 1.5 times service working load level (AS 2159 1995), and,
b. From two times service working load level (PWD, Singapore 1995; PWD Malaysia 1988). The magnitude of residual settlement upon unloading indicates the extent of yielding of the pile-soil interface interaction, particularly for long predominantly friction piles. For a given pile-soil system, the smaller the magnitude of residual settlement, the greater the extent of pile-soil

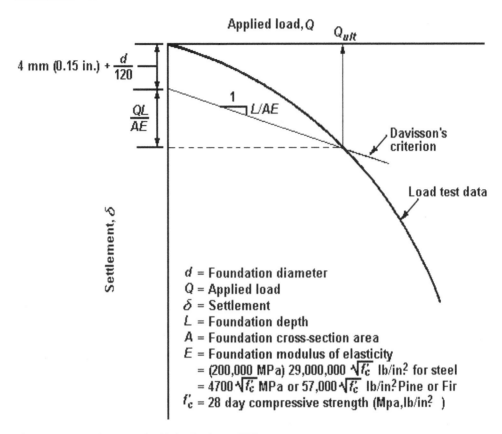

Figure 11.4.   Davisson's method (after Davisson 1973).

interface yielding that has occurred during the loading of the pile. As such, the magnitude of pile residual settlement is expected to increase as the ultimate geotechnical capacity of the pile is approached.

It is of interest to note that the BS 8004 and Euro code 7 do not provide specific recommendation for pile settlement acceptance criteria other than methods of estimating the ultimate geotechnical pile capacity, for example, capacity at 10% of pile diameter settlement (BS 8004, 1986).

The Australian standard (AS 2159 1995) provides specific settlement criteria of 15 mm and 50 mm at working load and 1.5 times working load respectively. Limiting residual settlements of 7 mm and 30 mm are specified for unloading from working load and 1.5 times working load respectively. AS 2159 further provides for no more than 3 mm creep to be included in the settlement and residual settlement values at working load level.

The Singapore (PWD 1995) specification, commonly referred to and adopted in Singapore, provides for a maximum pile settlement of 25 mm at a maximum test load of two times service working load for piles with a working load greater than 65 tons. Limiting pile residual settlements after unloading (from two times working load) of 5 mm and 0.05 mm per ton of working load (subject to maximum 13 mm) are specified for working loads less than or equal to 100 tons, and working load greater than 100 tons respectively.

The Malaysian PWD specification (1988) is referred to and routinely adopted by government agencies and the private sector for engineering projects in Malaysia, limits the pile settlement magnitudes of 12.5 mm and 38 mm (or 10% of pile diameter/width, whichever is the lower) and are specified at working load and two times working load respectively. The specification further

Table 11.1. Some relevant single pile settlement acceptance criteria.

Code/specifications	Pile settlement (mm) at		Pile residual settlement** (mm)
	Working load	Max. test load*	
AS 2159 (1995)	15.0 (see "Note 1")	50.0 (at 1.5 × WL)	= 7.0 mm (unloading from WL; see "Note 1"). = 30.0 mm (unloading from 1.5 × WL).
PWD, Singapore (1995)	–	= 13 mm for WL ≤ 65 tons. = 0.2 mm per ton for WL > 65 tons, but not more than 25 mm. (see "Note 2")	=5.0 mm for WL ≤ 100 tons = 0.05 mm per ton of WL for WL > 100 tons, but not more than 13 mm.
PWD, Malaysia (1988)	12.5	38.0 (or 10% of pile dia./width which-ever is the lower)	6.5

Notes: *: maximum test load normally taken as 2.0 times service working load (WL) unless otherwise specified. **: refers to unloading from two (2) times service working load unless otherwise specified.
[1]: Settlement to include no more than 3 mm creep over 5 hours after load has been in place for 15 mins.
[2]: For steel piles only, the maximum settlement under full test load given above may be increased by not more than 0.15 mm per meter length of the pile to allow for the elastic shortening of the piles.

provides for a single limiting residual settlement of 6.5 mm upon unloading from two times service working load irrespective of the pile size.

Table 11.1 summarizes some the pile settlement acceptance criteria. It appears from Table 11.1 that the settlement acceptance criteria do not specifically differentiate, for example, the types and sizes of piles, ground conditions such as mainly granular or clayey grounds, predominantly friction or end-bearing piles, and embedded pile lengths.

Chin (2004) presents results of a static load test on a 825 mm diameter drilled shaft and assessed it with regard to the relevant settlement acceptance criteria of Table 11.1. The Malaysian PWD specification for piling works was adopted as the standard specification governing the testing and foundation pile performance.

The 825 mm diameter test drilled shaft was installed to a length of about 10.3 m below the platform level. The drilled shaft was constructed dry through an upper layer of about 3.0 m of stiff clayey silt soil (average SPT N value of 12), followed by about 4.0 m of very stiff to hard clayey silt soil layer (average SPT N value of 40), and finally socketed about 3.3 m into a very hard clayey silt soil layer with SPT N values in excess of 200 blows per 300 mm penetration. The design working compression axial load for the drilled shaft was 270 tons, with maximum test load of 540 tons corresponding to two times the design working load. Figure 11.5 shows the field measured load-settlement response of the drilled shaft subjected to two cycles of static loading, (a) first cycle loading to design working load of 270 tons, and (b) second cycle to the maximum test load of 540 tons.

Table 11.2 tabulates the measured pile settlements and residual settlements at relevant static test load levels.

Table 11.3 tabulates the settlement performance acceptance of the test drilled shaft against the relevant acceptance criteria. The settlement acceptance criterion at maximum test load was apparently not satisfied for the Singapore PWD (1995) specification. The residual settlement criterion however was not satisfied by both the Singapore and Malaysia PWD specifications except for the AS 2159.

The measured maximum pile settlement was about 3.4% of the pile diameter, indicating possibly that the shaft capacity may have been fully mobilized with the end-bearing capacity being

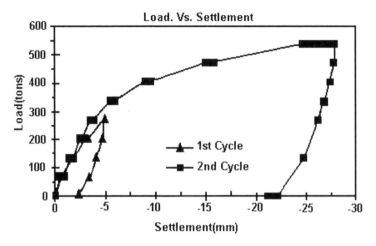

Figure 11.5.   Measured load-settlement response of the 825mm diameter test drilled shaft (after Chin 2004).

Table 11.2.   Measured settlement of the 825 mm diameter test drilled shaft.

Pile load (tons)	Pile settlement (mm)	Loading cycle stage
270	5.3	Cycle No.1 (100% WL)
0	2.2	
540	28.0	Cycle No.2 (200% WL)
0	21.2	

Table 11.3.   Pile performance acceptance based on relevant settlement acceptance criteria.

| Code/Specifications | Pile settlement performance at | | Pile residual settlement performance |
	Working load	Max. Test load	
AS 2159 (1995)	Satisfied	Satisfied (at 1.5 × WL)	Satisfied (both unloading from WL and 1.5 × WL)
PWD, Singapore (1995)	No settlement criterion provided	Not satisfied	Not satisfied
PWD, Malaysia (1988)	Satisfied	Satisfied	Not satisfied

progressively mobilized. An estimate of the ultimate pile capacity using Chin's (1970) method indicated an ultimate capacity of the order of about 640 tons, with a factor of safety of about 2.4 (= 640/270). A third cycle was subsequently conducted on the same pile and indicated a mobilized pile capacity of about 600 tons at a total pile settlement of about 45 mm (i.e. about 5.5% of pile diameter). In terms of pile geotechnical design capacity, the proposed pile length would have been considered quite an optimum design with a factor of safety of not less than 2.0. Chin (2004) suggested that the recommendations of AS 2159 (1995) are more consistent and practical as regards the residual settlement criterion, compared with either the Malaysia or Singapore PWD specifications.

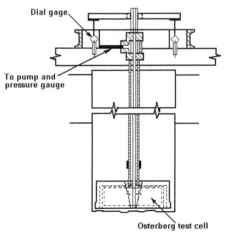

(a) Schematic diagram of Osterberg test cell

(b) Photograph of the Osterberg load test. (www.loadtestasia.com/about.htm)

Figure 11.6.   Osterberg load test device.

11.3   OSTERBERG LOAD TEST

Osterberg (1984) developed a method that reduces the cost of conducting high capacity static load tests, in particular for the drilled shafts, as shown in Figure 11.6. Once the concrete is in place, the operator pumps hydraulic fluid into the jack and keeps track of both pressure and volume. The jack expands and pushes up on the shaft. A dial gauge measures this movement, from which a plot of side friction capacity versus axial movement can be obtained. The devise also includes a telltale rod that extends from the bottom of the pancake jack to the ground surface. It measures the downward movement at the bottom, and thus produces a plot of toe-bearing pressure versus axial movement.

11.4   DYNAMIC METHODS

Methods to estimate axial pile load capacity based on pile driving formulae have been described in Chapter 5. In this section, we describe the dynamic methods that are based on the wave equation analyses.

Pile driving formulas are generally considered as unreliable. The alternative is to use the dynamics of pile driving in more detail, and this method is called the wave equation method. This method provides a more accurate function of capacity versus blow count; helps optimize the driving equipment, and computes driving stresses.

Pile-driving formulas consider the pile to be a rigid body subject to classical Newtonian physics. In other words, they assume the entire pile moves downward as a unit. In reality, the impact load provided by the pile hammer is very short compared to the time required for the resulting stress wave to reach the bottom of the pile; so portions of the pile may be moving downward, while other portions are stationary or even moving upward (in response to a reflected wave). Therefore, it is much better to consider stress wave propagation effects when evaluating the pile driving process.

In 1976, researchers at the Case Institute of Technology developed the WEAP (Wave Equation Analysis of Piles) program. It has been revised on several occasions and is in public domain (http://uftrc.ce.ufl.edu/info-cen/info-cen.htm). The WEAP program has since formed the basis for another advanced proprietary program, the GRLWEAP (www.pile.com). Wave equation analysis software is widely available, and can easily be run on personal computers.

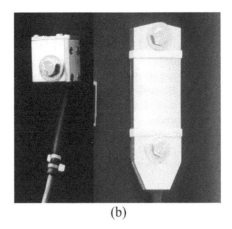

(a)   (b)

Figure 11.7.   (a) Pile-driving analyzer; (b) Accelerometer (left) and strain gage (right) mounted near the top of a pile to provide input to the pile-driving analyzer (Pile Dynamics Inc., http://web.pile.com/pdi/default.asp?Company=PDI).

Another dynamic method of evaluating the static load capacity of deep foundations is to install instruments on the foundation and use them to monitor load and settlement data obtained while the foundation is subjected to a dynamic impact load. This measured response to dynamic loads can then be used to develop design static load capacities.

The most common source of dynamic loading is a pile hammer, because it is already on site and thus represents little or no additional cost. Therefore, these tests are most commonly performed on driven pile foundations. However, dynamic loads also can be obtained with drop hammers that enable testing of drilled shafts and other types of deep foundations.

This method requires sufficient strain in the foundation to mobilize the side friction and toe bearing, and thus is called high-strain dynamic testing. There is also a low strain dynamic testing, which is used to evaluate structural integrity.

Field equipment for measuring the forces and accelerations in a pile during driving was developed during the 1970s and are now commercially available.

This equipment includes three components:

– A pair of strain transducers mounted near the top of the pile.
– A pair of accelerometers mounted near the top of the pile.
– A pile driving analyzer (PDA).

The strain transducers and accelerometers and a pile driving analyzer is shown in Figure 11.7.

The pile driving analyzer monitors the output from the strain transducers and accelerometers as the pile is being driven, and evaluates this data as follows:

– The strain data, combined with the modulus of elasticity and cross-sectional area of the pile, gives the axial force in the pile.
– The acceleration data integrated with time produces the particle velocity of the waves traveling through the pile.
– The acceleration data, double integrated with time produces the pile displacement during the hammer blow.

Using this data, the PDA computes the Case method capacity, and displays the results immediately. It also can store the field data on a floppy disk to provide input for a CAPWAP analysis.

Case method analyses

The Case method is an analytical technique for determining the static pile capacity from wave trace data (Hannigan 1990). The PDA is programmed to solve for pile capacity using this method and gives the results of this computation in real time in the field.

This Case method computations include an empirical correlation factor, j_c, that can be determined from the static load test. Thus, engineers can use this method to extend static load test results to selected production piles. However, for most projects, it would not be cost-effective to obtain PDA measurements on all of the production piles.

It also possible to use the Case method without an on-site pile load test by using j_c values from other similar soils. This approach is less accurate, but can still be very valuable.

CAPWAP

The Case method, while useful, is a simplification of the true dynamics of pile driving and the associated response of the adjacent soil. The empirically obtained damping factor, j_c, calibrates the analysis, so the final results are no better than the engineer's ability to select the proper value. In contrast, a wave equation analysis utilizes a much more precise numerical model, but suffers from weak estimates of the actual energy delivered by the hammer. However these two-combined methods are to form an improved analysis known as the CAPWAP (Case Pile Wave Analysis Program) (Rausche et al. 1972).

A CAPWAP analysis performed on PDA data could be used as follows:

– To provide more accurate input parameters for a wave equation analysis that could then be used to select the optimal driving equipment as well as to produce a bearing graph.
– To provide a site-specific Case method damping factor, j_c, for use in PDA analyses of selected production piles.
– To obtain quantitative measurements of pile setup (Fellenius et al. 1989).
– To produce simulated static load test results.

CAPWAP analyses can be used to reduce the required number of static load tests, or used where load tests are not cost-effective.

Statnamic test

The statnamic test is another high strain dynamic testing method for evaluating the static load capacity of deep foundations. This method loads the foundation by detonating slow-burning explosives located inside a pressure chamber placed between the foundation and a mass, as shown in Figure 11.8. The force from the explosion generates a downward movement in the foundation, which is monitored using load and displacement instruments, and the data obtained from these instruments is plotted directly as a load-displacement curve similar to that obtained from a static load test. The mass provides a reaction for this force.

Comparison of test results of high strain dynamic testing with static load test

Figure 11.9 shows the comparison of interpreted pile axial capacity of driven precast spun and square piles from the static load test (SLT) and high strain dynamic pile testing (HSDPT) from several project sites in Malaysia. With some exceptional test results, the variation of the predicted pile capacity generally falls within 15% for the test piles loaded to ultimate state or failure. The variation in over estimation of pile capacity is probably due to the inherent dynamic damping effect in the high strain dynamic testing. Rausche et al. (1985) reported that any value of damping constant between 0.0 to 2.0 would gave results within 20% of the statically measured value. The prediction of pile capacity is usually sensitive to the selection of damping constant unless the pile toe velocity during driving is zero.

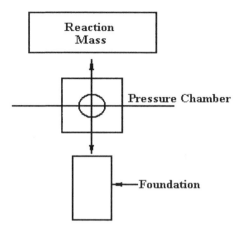

Figure 11.8.   Schematic of the statnamic test.

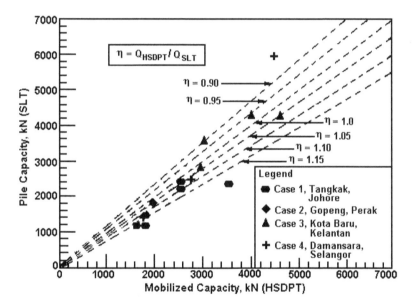

Figure 11.9.   Comparison of pile axial load capacity from static load test (SLT) and high strain dynamic pile testing (HSDPT) (from Liew et al. 2004).

11.5   CONCRETE PILE INTEGRITY

The integrity or quality of a grouted concrete pile can be checked using the acoustic wave method, such as the PIT (Pile Integrity Tester), which is a low stain dynamic testing, or the ground penetrating radar (GPR) technique.

Low-strain dynamic testing consists of striking the top of the foundation with a small load and monitoring the resulting waves using one or more accelerometers. These tests typically use a carpenter's mallet to generate the load, so the induced strains are very small and the settlement is not nearly sufficient to fully mobilize the side-friction or toe-bearing resistance. Low-strain dynamic tests do not provide an indication of static load capacity, but they are very useful for evaluating

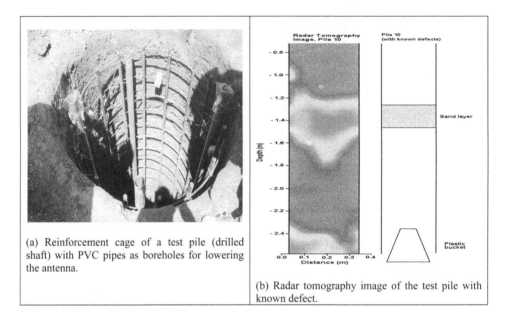

(a) Reinforcement cage of a test pile (drilled shaft) with PVC pipes as boreholes for lowering the antenna.

(b) Radar tomography image of the test pile with known defect.

Figure 11.10.   Checking concrete pile integrity using GPR technique (from Kong et al. 1997).

structural integrity (such as major cracks, necking, soil inclusions or voids), and for determining the as-built length of the pile.

In the GPR technique, measuring the wave transmissions while varying the transmitter antenna location in one borehole and the receiver antenna location in another borehole specially cast along the sides of a grouted concrete pile can perform tomography tests. The arrival and magnitude of the transmission wave will change when defects are met, which is used as input for the tomographic inversion, as illustrated in Figure 11.10.

REFERENCES

AS 2159 1995. Piling: Design and installation. Standard Association of Australia.

BS 8004 1986. Code of practice for foundations. British Standard Institutions. London.

Chin, F.K. 1970. Estimation of the ultimate load of piles from tests not carried to failure. *Proceedings of 2nd South East Asian Conference on Soil Engineering.* Singapore: 81–90.

Chin, J.T. 2004. A review of some pile settlement acceptance criteria. *Malaysian Geotechnical Conference. The Institution of Engineers*, Malaysia. Kuala Lumpur: 269–275.

Coduto, D.P. 2001. *Foundation design, principles and practices.* New York: Prentice Hall.

Davisson, M.T. 1973. High capacity piles. In innovations in foundation construction. *Proceedings of Lecture Series, Illinois Section, ASCE,* Chicago.

Eurocode 7 (ENV 1977-1) 1994. Geotechnical Design: Part 1 – General Rules, European Committee for Standardization.

Fellenius, B.H., Riker, R.E., O'Brien, A.J. & Tracy, G.R. 1989. Dynamic and static testing in soil exhibiting set-up., *ASCE Journal of Geotechnical Engineering.* 115(7): 984–1001.

Hannigan, P.J. 1990. Dynamic monitoring and analysis of pile foundation installations. Deep Foundations Institute, Sparta, NJ.

Kong, F.N., Bahia, H.M. & Huat, B.B.K. 1997. Radar tomography checking concrete pile integrity. *Non Destructive Testing.* BSI, Liverpool, UK.

Liew, S.S, Wg, H.B. & Lee, K.K. 2004. Comparisons on HSPDT and SLT results on driven piles in Malaysia residual soils. *Malaysian Geotechnical Conference.* The Institution of Engineers, Malaysia. Kuala Lumpur: 315–321.

Osterberg, J.O. 1984. A new simplified method for load testing drilled shafts. Foundation drilling. Association of drilled shaft contractors. Dallas.

PWD 1988. Specification for piling works. Public Works Department, Malaysia.

PWD 1995. General specification: Piling. Public Works Department, Singapore.

Rausche, F., Mosses, F. & Goble, G. 1972. Soil resistance predictions from pile dynamics. *ASCE Journal of the Soil Mechanics and Foundation Division*. 98. SM9.

Rausche, F., Goble, G. & Likins, G.E. Jnr. 1985. Dynamics determination of pile capacity. *Journal of Geotechnical Engineering, ASCE* 111(3): 367–383.

CHAPTER 12

Special Topic: Foundation Failures and Remedial Works

Bujang B.K. Huat

Department of Civil Engineering, University Putra Malaysia, Malaysia

12.1 INTRODUCTION

Foundations are designed to satisfy at least two main criteria:

1. Adequate stability, i.e. the factor of safety with respect to load carrying capacity should be adequate.
2. The settlement should be tolerable with respect to the structure. These two criteria apply to all foundations.

Occasionally, the foundations of engineering structures do not perform as intended. When a foundation failure occurs, it is important to study it carefully to see what has gone wrong. For the profession as a whole, this knowledge is useful to prevent a recurrence. The challenge to the engineering profession is to reduce recurrences of past failure to a minimum. From the point of view of engineering science, a foundation failure is a destructive full-scale experiment up to the ultimate state. As such, useful lessons can be learnt. On the other hand, a failure is also a painful and expensive experience for the parties involved, and for society as a whole. Therefore since the price is already paid after a failure, it is important to learn from the experience (Chan 1996, 2004).

Compared with the superstructure, foundations can be considered to be more complex, the reasons for which can be attributed as follows. In a superstructure the main elements, i.e. concrete and steel, and the design and construction processes are all man-made. Since quality control can be brought to bear on these man-made materials, the degree of uncertainty associated with these materials is relatively low and easy to control. The foundation on the other hand, besides the concrete and steel, there is a new predominant element, i.e. the ground, which is crucial in both the design and construction processes. Being formed by nature, the ground is characterized by heterogeneity (non-uniformity), great variation in engineering properties and complexity in behavior. For example, even within the same project site, the strength of the ground can differ by a factor of 10 to 100, compared with the usual factor of safety for stability analysis of only 2 to 4. All these have an important impact on the safety of foundations.

In view of the above, no two foundations are exactly alike. Terzaghi (1936) had realized several decades ago that the ground is full of surprises for the unwary. In spite of modern advances in site investigation and geology, this statement is still true today. Peck (1988) pointed out that every construction job involving earth or rock runs the risk of encountering surprises. Thus the complexity of foundations stems largely from the complexity of the ground. Any inadequacy in dealing with the ground can lead to foundation difficulty or failure.

Another factor which has exacerbated the problem is that in some foundations, e.g. drilled shafts (or bored piles) and caisson, part of the construction process, take place out of sight below ground. Furthermore, for all pile foundations, the pile shafts below ground are not visible or accessible after completion. Such lack of visibility hamper quality control although special measures, such as sonic logging, may be adopted to alleviate the problem.

As in all geotechnical structures such as dams, earth retaining structures and foundations, in addition to stability, deformation or consolidation and time also require major consideration. Distribution of forces and load, and hence stability, can be affected by the deformation. Common

Table 12.1. Soil profile: Case 1.

Depth (m)	Description
0–21	Very soft to soft marine clay
21–27	Medium stiff clay
27–50	Medium dense to dense sand

examples are downdrag on piles and redistribution of pile loads due to deformation of the pile raft. Another example which has caused many failures is the application of horizontal loads on piles due to surcharge loading on the ground nearby, e.g. from floor load or from weight of embankment. Furthermore, since deformation is time dependent, stability can also be time dependent. Unfortunately, the methods for determining deformation and effects of time are more complicated and fraught with uncertainties compared with those for superstructures, since ground properties and behavior feature prominently in these methods. In view of the above factors, it is not surprising that occasionally foundations do not perform as intended, and as a result foundation failures occur.

12.2   SOME CASE HISTORIES OF FOUNDATION FAILURES IN TROPICAL SOILS

A foundation failure provides a rare opportunity to study the fundamental behavior of foundations up to the ultimate state, thereby improving the state of the knowledge of foundations. Such knowledge is the basis for the design of foundations. Furthermore, before attempting any remedial work, it is essential to understand the behavior of the foundation which is undergoing failure. However, this is a complex matter. This is because during failure the behavior of the foundation with respect to movement, time, and location is governed not only by the type of foundation but also by ground conditions and the characteristics of the superstructure. In addition, the interaction of these three factors also has a strong influence on the behavior, e.g. redistribution of loads. For the reasons mentioned above, every foundation failure must be regarded as unique. Clearly every case will have to be studied individually. The importance of understanding the behavior of mechanism that contributed to the failure has been emphasized by Burland (2004).

Case 1: An oil storage tank

A steel oil storage tank, 17.0 m diameter and 15.0 m high, was built on the coastal alluvium of the West Coast of Peninsular Malaysia. A simplified subsoil profile is shown in Table 12.1.

The tank rested on a sand pad laid on a thin R.C. raft, 0.22 m thick, which was in turn supported uniformly on a total of 1250 *bakau* (mangrove) timber piles. Each pile, which was 100 mm diameter, was 20 m long with 3 sleeve joints and a 2 m long R.C. dolly at the top. The piles were installed at 0.5 m apart. The design fluid for the tank had a specific gravity of 0.87.

The tank was load tested by filling gradually with water. When the water level reached a height of 13.2 m, the tank failed by tilting to one side, accompanied by excessive settlement of up to 0.89 m. After the failure, the tank was propped up by jacking to facilitate examination of the R.C. raft.

Figure 12.1 shows the deflected shape of the raft in the direction of the tilt. Due to the excessive total and differential settlement, there was extensive structural damage in the form of wide cracks and a rupture occurred in the raft directly beneath the vertical tank shell on the side of the tilt. The deformation of the tank base was such that the tank superstructure was at a high risk of rupture.

The cause of the failure was insufficient depth of penetration of the piles. There was no site investigation specifically for the tank, only reliance on geology of the site and past experience with similar tanks (Chan 1996).

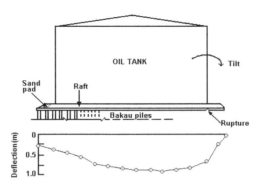

Figure 12.1.   Deflected shape of raft (Chan 1996).

Table 12.2.   Soil profile: Case 2.

Depth (m)	Description
0–1	Fill
1–18	Very soft to soft marine clay
18–25	Medium dense sand
Below 25	Sandstone

Case 2: A row of 2-storey shops

A row of seven shop houses in the West Coast of Peninsular Malaysia located on subsoil with a profile as summarized in Table 12.2.

The foundation consisted of 100 mm diameter *bakau* (timber) piles with a nominal working load of 1 ton per pile. All the piles were driven to a depth of 7 m only with pile caps at 1 m depth. The pile toe level was still well within the soft clay. This was the cause of the foundation failure.

When the buildings were near completion, signs of damage became apparent. Firstly, cracks appeared on the brick walls and soon the first floor R.C. slab also suffered cracking. The cracks continued to widen in spite of repeated patching. Investigations and detailed monitoring of settlement started 21 months after the buildings were completed. Monitoring of the settlement of the columns was carried out using precise leveling technique to an accuracy of 0.2 mm.

When time detail monitoring was started, the maximum total settlement of the row of seven buildings was estimated to be at least 250 mm. The settlement was far from uniform. By visual inspection, it appears that the settlement profile was dish-shaped. The settlement at the centre had exceeded the average settlement of the two ends by 149 mm.

Each of the seven buildings had undergone substantial differential settlement, ΔS, and angular distortion, $\Delta S/L$, where L is the distance between adjacent columns. Figure 12.2(a) shows the magnitude of differential settlement and angular distortion for each building. For comparison, the limiting angular distortions of Bjerrum (1963) which mark the onset of serious structural damage to various types of structures as shown in Figure 12.3, is referred. The buildings near the two ends had undergone angular distortions exceeding the limiting value of 1/150, which marks the onset of serious structural damage to R.C. frame buildings.

The settlement behavior with respect to time is shown in Figure 12.4. The entire row of buildings continued to sink into the ground with the middle of the row settling at a faster rate (5.9 mm/month) compared with the two ends (3.7 and 4.5 mm/month). Thus the dish-shaped settlement profile was becoming more accentuated with time. This means that differential settlement, and hence structural damage, were increasing with time.

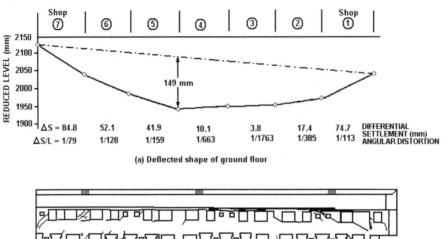

(a) Deflected shape of ground floor

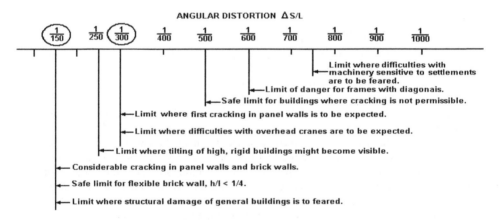

(b) Crack pattern at rear of shops

Figure 12.2.   State of buildings 21 months after completion (Chan 1996).

Figure 12.3.   Limiting angular distortions (Bjerrum 1963).

Case 4: An R.C. bridge foundation

The project was a reinforced concrete (R.C.) bridge over a river in Selangor, Malaysia. The proposed heights of the approach embankments on both sides of the abutments were about 8 m with side slope of $1V$ (vertical) to $1.5H$ (horizontal). These embankments were to be constructed over a layer of very soft silty clay or clayey silt of 3 m to 9 m thick with Standard Penetration Tests values (SPT) N of zero. Underlying the very soft top layer was 3.5 m to 5.5 m thick of medium dense silty sand followed by completely weathered shale with SPT 'N' values varying between 30 to 50 blows/300 mm. The liquid limit (w_l) of the clay was about 78% and average moisture content was about 106%.

Figure 12.5 shows the general layout of the project. Figure 12.6 shows a schematic profile of the subsoil. At Abutment A, the wing walls were designed on piles but Abutment B was designed with cantilever wing walls.

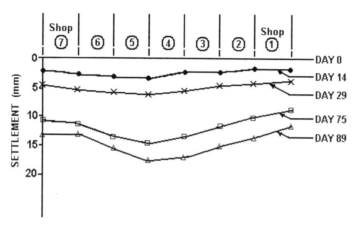

Figure 12.4. Settlement-time behavior starting 21 months after completion (Chan 1996).

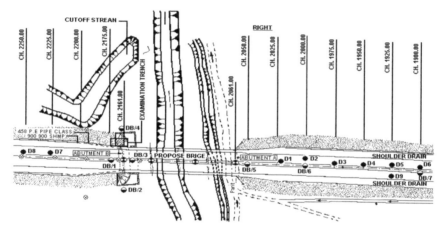

Figure 12.5. Layout of the R.C. bridge over a river in Selangor, Malaysia (Gue & Tan 2003).

The first slip failure occurred on the right side of the approach embankment behind Abutment B when the fill was about 3 m high above ground level. The designer inspected the failure and gave instructions to build a berm and continue filling. Three days later, the second slip failure at the site occurred. This time, it was on the left side of the same embankment as shown in Figure 12.7 (a). Despite the failure, the designer instructed the contractor to dig a ditch of 0.5 m wide by 1 m deep at the toe of left side of the embankment and allowed earth-filling works to continue towards Abutment B at a slower rate. Piling works at Abutment B started about 6 weeks after the first slip and the construction of the whole abutment was completed in about 5 weeks. After two months and when the embankment reached a height of 7 m (1 m below formation level), the third slip failure occurred as shown in Figure 12.7 (b). The third slip occurred on the same location as the first slip and together with the embankment directly behind Abutment B. A drop of about 2.4 m was observed at the slip interface. Abutment B was found to have moved forward by 0.31 m and 1.12 m on the left and right sides respectively of the abutment. The abutment had also tilted vertically about 2° clockwise towards the embankment.

In the failure investigation of Abutment B, the piles immediately below the pile cap were examined by excavating a trench on the right side of the Abutment B. Figure 12.8 and 12.9 show the condition of the piles. Cracks on pile B appear to have propagated from the left side of the pile

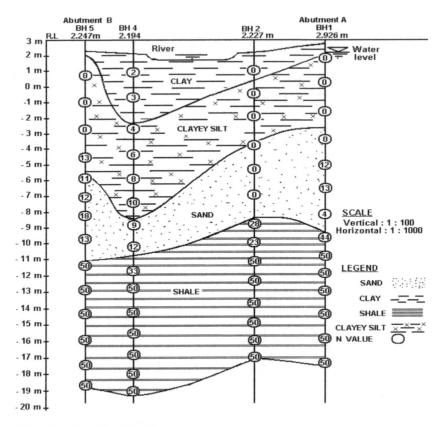

Figure 12.6.   Subsoil profile of the site.

Figure 12.7.   (a) Second slip failure (left); (b) Third slip failure.

towards the fill. This indicates that the piles had been subjected to high lateral stresses imposed by the fill. The pile also showed a slight curve that suggested that there had been some restrain at the lower end of the pile.

Figure 12.10 shows the crushing of Pile C, which further indicates lateral compression due to the fill. Figure 12.11 shows schematically the movements of pile cap and piles.

Figure 12.8.   View of exposed pile cap and piles on the right side of Abutment B.

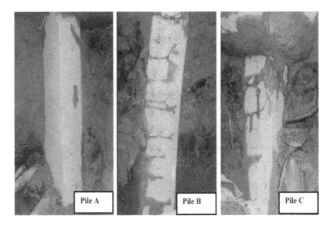

Figure 12.9.   Close-up view of the three exposed piles.

Figure 12.10.   Pile C showing crushing of pile near pile cap.

The fourth slip occurred on the left side of the embankment, this time behind Abutment A, about three months after the third slip. At the time of the failure, the height of the embankment was about 6.5 m above its ground level (1.5 m below proposed formation level). Figure 12.12 shows the slip failure.

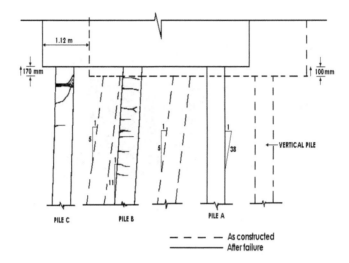

Figure 12.11.   Movements of pile cap and piles (view from right side of Abutment B).

Figure 12.12.   Forth slip failure on the left side of embankment behind Abutment A.

In order to ascertain the causes of the failures, geotechnical analyses and investigations were carried out. According to Gue & Tan (2003), results of the soil investigation generally confirmed the results of the previous soil investigation except that the very soft silty clay layer appeared to have gained some strength and decreased in thickness due to some consolidation.

Generally, the average undrained shear strength (S_u) of the very soft clay was 10 kPa with a lower bound value of 7.5 kPa as shown in Figure 12.13. The sensitivity (S_t) of the clay ranged from 2 to 8.

Bearing capacity and limit equilibrium stability analyses carried out indicate that the subsoil could not support embankment height in excess of 2.7 m without any ground treatment or strengthening. From the analyses, it appears that the embankment height of 8 m proposed by the designer was not safe. Further back-analyses carried out on the failed embankment indicated that the S_u of the clay was 11 kPa. This was in good agreement with the average S_u of 10 kPa obtained from the site investigation.

Analyses on the foundations for Abutments A and B were also carried out. The lateral earth pressure on piles was calculated using stress distribution behind piles proposed by Tschebotarioff (1973) as shown in Figure 12.14. Since the spacing of piles was about three times the width of the pile, the group of piles and soil can be assumed to act as a unit. The ultimate lateral resistance, R_u was calculated using assumption of Poulos & Davis (1980) that assumed $4S_u$ at the surface, increasing to a constant value of $9S_u$ at three times the width of the pile. The critical height of

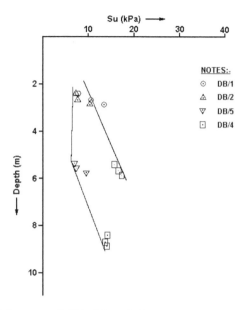

Figure 12.13.   Undrained shear strength (S_u) of the soft clay.

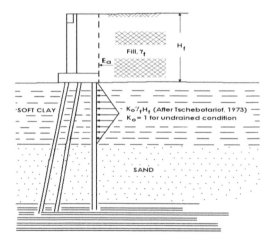

Figure 12.14.   Additional forces on abutment (soil undergoing lateral movement).

embankment that would induce a lateral force equivalent to the ultimate resistance was evaluated for the different thickness of soft clay.

For Abutment B, the lateral resistance of the pile group and soil is exceeded when the height of embankment is 5 m to 5.5 m. Therefore, the movements of Abutment B could have happened when the height of embankment was about 5 m, i.e. 3 m below the proposed formation level. The calculations had also shown that for Abutment A, the lateral resistance of the pile group and soil would be exceeded when the height of embankment was 7.5 m to 8.0 m. However, Abutment A was only subjected to a height of 6.5 m (1.5 m below the proposed formation level), therefore it did not fail. Abutment A could withstand a higher embankment height than Abutment B because of its shape and dimensions. Abutment A was designed and constructed with piled wing-wall. The total number of piles used was 88, and had 36 piles more than Abutment B.

Figure 12.15.   Overview of partially completed bridge in Sarawak, Malaysia (Gue & Tan 2003).

Finally, the proposals for the remedial works included the following:

a. Underpinning of Abutment B with 48 micropiles, Abutment A with 10 micropiles.
b. Use of reinforced soil wall behind Abutment A.
c. Piled embankment to be used for embankment fill exceeding 2.5 m high.
d. Geogrid reinforced embankment for fill between 1.5 m to 2.4 m high.

The total estimated cost of remedial works was about 3.7 million ringgit (about USD 0.97 million) (Gue & Tan 2003).

Case 5. A pre-stressed bridge foundation

The project, a pre-stressed concrete bridge over a river in Sarawak, Malaysia, was under construction when failure occurred. The proposed heights of the approach embankments on both sides of the abutments were about 5 m with side slopes of $1V$ (vertical) to $1.5H$ (horizontal). The embankments were constructed over 25 m thick soft coastal and riverine alluvium clay followed by dense silty sand and very stiff silty clay. Soft alluvium generally has SPT, N, value of zero and average moisture content of more than 70%. Figure 12.15 shows the partially completed bridge after failure and removal of the failed materials. The layout is shown in Figure 12.16 and the subsoil profile in shown in Figure 12.17.

In the construction drawings, the approach embankments using local fills were to be supported by 200 mm × 200 mm RC piles with pile-caps. In addition, 6 m long wooden piles were also used between the RC piles for further support of the embankment fill. More wooden piles were also installed on the banks of the river in an attempt to stabilize the lateral displacement of the soft alluvium. The abutments and piers were generally supported by 400 mm diameter spun concrete piles.

A deep-seated slip failure occurred at the approach embankment about 25 m from Abutment II. It happened when the fill reached about 3 m high. Figure 12.18 shows the shear drop after removal of some of the fill near the abutment.

Abutment II has tilted away from the river with a magnitude of about 550 mm at the top of the abutment at the time of the site inspection by Gue and Tan (2003) who were carrying out geotechnical investigation of the failure. The tilt translated into an angular distortion of 1/6. Due to the excessive angular distortion, the integrity of the spun piles driven to set into the stiffer stratum had also been affected as it exceeded the normal threshold of about 1/75. Due to the tilt of the Abutment II away from Pier II, a gap of about 300 mm wide was observed between the two bridge

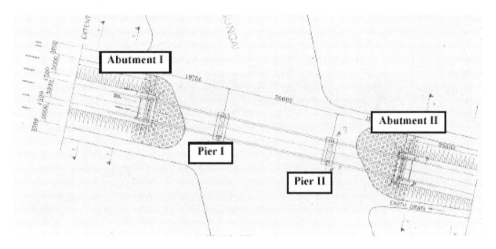

Figure 12.16.   Layout of the pre-stressed bridge.

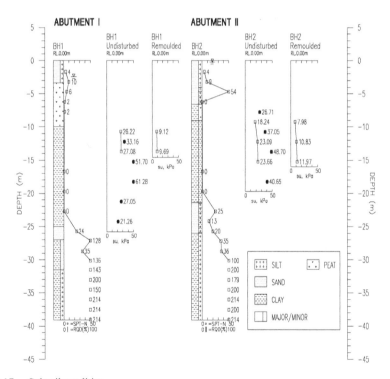

Figure 12.17.   Subsoil condition.

decks at the pier's pile cap. Figure 12.19 shows the photograph of the tilt at Abutment II and the gap between two bridge decks.

The failure also caused the pile cap at Pier II to tilt as shown in Figure 12.20. Figure 12.21 shows the schematic diagram of the possible slip plane relative to the deformed structures.

These observations infer that the slip of the Approach Embankment near Abutment II was deep seated and was consistent with the depth of the soft alluvium. Gue and Tan (2003) postulated that the cause of the rotational slip failure was due to the weak subsoil being unable to support the

Figure 12.18.   Shear drop at about 25 m from tilted abutment.

Figure 12.19.   Tilted abutment and observed gap between bridge decks.

Figure 12.20.   Titled pilecap at Pier II.

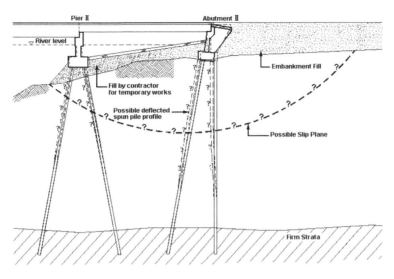

Figure 12.21.   Schematic of the slip failure.

weight of the approach embankment. The weight of embankment initiated the consolidation settlement of the soft subsoil and mobilized the low shear strength of the slip failure plane. The use of the 6 m wooden piles and RC piles offered little lateral resistance and instead, extended the rotational slip deeper into the soft subsoil. At the pier, the bridge decks, being simply supported and fixed to the abutment via bearing pad, had moved along with the displacement of the abutment.

The upper subsoil had an undrained shear strength of about 10 kPa. Using correlation of $5 \times S_u$ to obtain ultimate bearing capacity of 50 to 60 kPa, the estimated maximum height of fill that could be supported was about 3 m which was consistent with the observed failure during embankment filling to 3 m high.

Additional subsurface investigation after the failure showed that the undrained shear strength from the vane shear tests ranged from 18 kPa to 51 kPa with remolded strength of 7 kPa to 12 kPa. The higher S_u obtained from the additional site investigation was probably due to the gain in strength during the whole period of filling.

According to Gue & Tan (2003), the rotational slip failure of the approach embankment was due to many factors such as inadequacy in design, construction control, the soft subsoil and absence of adequate ground treatment.

Several remedial options were explored for the embankment. The first remedial option was to remove the failed embankment fill and re-construct a new RC ramp (bridge) with ground beams for increased rigidity. This option avoids the weight of the fill bearing on the soft subsoil. The second option was to surcharge the soft subsoil in combination with prefabricated vertical drains to accelerate the consolidation process of the clayey subsoil and allow the subsoil to gain strength with time. The third option was to use piled embankment with slab to transfer the embankment load to the stiffer soil stratum instead of the soft upper clay. The third option was finally chosen for it offered the shortest construction time in order to put the bridge into service with no long-term risk of further subsoil settlement. However, in this option, the soffit of the RC slab should be at or below the original ground level to avoid additional load on the soft upper subsoil stratum that can generate negative skin friction on both the abutment and piles.

At the tilted abutment, analyses of the pile head movement of the existing piles showed that integrity of the piles was doubtful and should be compensated. There were two options for installing the compensation piles; firstly at the sides of the existing pile group and secondly, behind the abutment. The first option requires demolishing the existing abutment and enlarging the pile-cap. The second option minimizes modification of the abutment but requires longer I-beams for support

Figure 12.22.   Cracked due to continued settlement (Chan 1996).

of the bridge deck. In addition, there was also a risk the compensation piles might be impeded by the wooden and RC piles since the location of these piles might have been displaced along with the slip failure. The first option was finally chosen to minimize remedial cost by reusing the existing I-beams and minimizing risk of the new spun piles striking the existing wooden and RC piles. For the pier foundation, the existing spun piles were fully compensated by demolishing the existing pier and pile-cap with installation of new ones at the sides. The total estimated cost of remedial works was about 1.3 million ringgit (USD 0.34 million).

Case 5: Three-storey shops

The three-storey shops in Shah Alam, Selangor, Malaysia were supported on piles but the 2 m wide walkway rested directly on a 3 m thick layer of poorly compacted residual soil fill, underlain by 5 m thick soft marine clay.

The building was intact but the settlement of the walkway continued causing it to crack as shown in Figure 12.22.

In this connection it is worth noting some of the common observations on settlement of poorly compacted residual soil fills, especially those with substantial silt and clay size particles. The settlement, whether due to self-weight or applied load, is invariably time-dependent, extending to many years or even decades, particularly for deep fills. This is in contrast to the common perception that settlement under self-weight will cease after a year or two. Moreover, the magnitude of settlement under self-weight alone can be very considerable, e.g. settlement up to 3% or 4% of thickness of fill within the first few years is not uncommon (Chan 1996). Tropical residual soils are also susceptible to large sudden settlement upon saturation by rising ground water level (Ting 1979). This phenomenon is referred to as "collapse settlement". We now know this to be due to loss of matric suction of a partially saturated residual soil upon saturation or rise of water table.

12.3   REMEDIAL WORKS

When a foundation failure occurs, several options are available.

1. To underpin the foundation thereby stopping the excessive settlement permanently.
2. To demolish the structure.
3. To leave the structure as it is.
4. Finally, grouting can be considered in appropriate circumstances.

However the third option is seldom viable because the excessive settlement which often tends to increase with time usually leads to increasing structural damage. This may bring about the collapse of the structure.

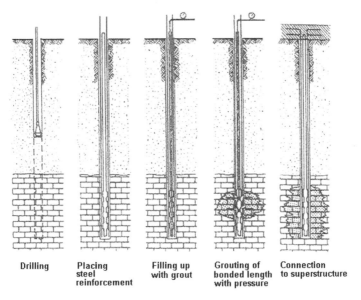

Drilling   Placing   Filling up   Grouting of   Connection
            steel   with grout   bonded length   to superstructure
            reinforcement   with pressure

(a) Construction Procedure

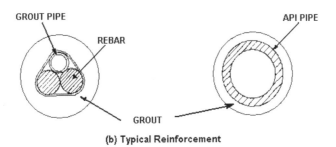

(b) Typical Reinforcement

Figure 12.23.   Micropiles (GEO 1999).

Examples of the demolition option that has been carried out in Malaysia are the demolition of the 16-storey Northam Court building in Penang (*The Star* 1980), and a row of 2-storey low cost flats in Sg. Way, Selangor (*The Star* 1983).

For the case of underpinning, one of the most widely used technique for underpinning foundation failures is the micropile (Chan 1996). Micropiles, sometimes known as mini piles (FPS 1987; GEO 1994), have been in use in various countries for five decades after it was first introduced in Italy in the early 1950s by Fernando Lizzi who called it root piles (Lizzie 1982; Stocker 1994; Bruce et al. 1995). In comparison with normal driven or bored piles, micropiles are smaller in diameter but are very heavily reinforced. For this reason the behavior of micropiles is akin to that of steel piles. The usual nominal overall size of pile ranges from 125 to 300 mm diameter. The working load per pile is typically 15 to 250 tons. Micropiles to a depth of 40 m have been constructed in Penang, Malaysia (Chan 1996). The construction procedure is shown in Figure 12.23. A micropile is constructed by first drilling a hole to the required depth using the rotary drilling method, except in very hard formations where down-the-hole hammers may be used. Bentonite slurry is commonly used to stabilize the borehole. An alternative is to use temporary casing. Figure 12.24 shows an auger drill for micropile.

When the final depth is reached, the reinforcement is lowered into the hole. Cement grouting of the hole then commences, starting from the bottom. In most ground conditions, pressure grouting is preferred.

Figure 12.24.   Photograph showing auger drill for micro-piles.

Figure 12.25.   Minarets of Masjid Jamek, Kuala Lumpur.

Two types of reinforcement are commonly used. The first is a very thick steel pipe which is preferred for very soft ground. This pipe can also serve as the "tube a manchettes". The second type of reinforcement is in the form of extra-large diameter steel bars. Cross-sections of typical micropiles with the two types of reinforcement are shown in Figure 12.23.

An example of successful underpinning work using micropiles is the rehabilitation of a mosque minaret describe by Chan (1996). Masjid Jamek is an old historic mosque situated right in the centre of Kuala Lumpur. It has two minarets, each with a height of 25.3 m and a octagonal base 2.8 m across (Figure 12.25). When it was discovered that one of the minarets was undergoing significant

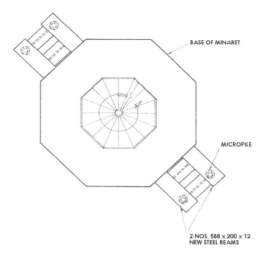

Figure 12.26.   Underpinning of minaret.

uneven settlement thereby developing a tilt, the solution adopted was to arrest the settlement by underpinning using micropiles.

The location site of the mosque is within the Kenny Hill Formation (quartzite and phyllite) of Kuala Lumpur, Malaysia. Inspection trenches excavated for examining the existing foundation and deep soil investigation boreholes revealed that the top 5 m was loose silty sand underlain by 3 m of medium dense sand. Below this were residual soils followed by weathered quartzite and phyllite of the Kenny Hill Formation.

The total weight of the minaret, which was estimated to be 100 tons, was supported on an old brick footing 2.1 m deep. In the remedial scheme, the entire load of 100 tons was supported on 4 micropiles as shown in Figure 12.26. Each micropile, which was 16 m long, was reinforced with a thick wall, high tensile API steel pipe.

In order to transfer the load to the piles, two steel transfer beams were inserted through the base of the minaret at ground level. The two beams were structurally connected together to enhance stability (Figure 12.26). Inserting steel rods and grouting reinforced the old brick footing.

Another possible option, i.e. in a relevant situation was to use permeation grouting which means pressure grouting with the aim of filling the pores between soil particles. The grout used is either (i) a particulate suspension (i.e. a suspension in water of cement, bentonite or p.f.a.), or (ii) a chemical gel, e.g. sodium silicate.

In Malaysia, permeation grouting mainly using cement is sometimes attempted as a solution to foundation failures, but with little success (Chan 1996, 2004). The reason for its failure was attributed to the following. The residual soils commonly used as fills have substantial proportions of clay and silt size particle. While grouting may fill up some of the large open voids in a poorly compacted fill, the fill remains largely unchanged. Thus settlement of the fill due to self-weight and imposed load will continue. Besides, the strength and compressibility of the soft soils beneath the fill cannot be improved by permeation grouting (Greenwood 1987).

12.4   CAUSES OF FAILURE AND LESSONS LEARNED

It is a sobering thought that foundation failures continue to occur in spite of rapid technological advances. In many instances this is because although state-of-the-art technology is available, to an individual practitioner it may not. In some other instances, the technology is available but pressure of time, budget and commercial considerations may prevent the complete utilization of current technology. In the vast majority of cases, the reason for the causes of foundation failures is human shortcomings, and these failures are largely avoidable. In a survey of failures of earth dams in North America, Peck (1981) concluded that nine out of ten recent failures occurred not because of inadequacies of technology, but because of the oversights that could and should have been avoided. Chan (1996, 2004) listed six main causes that can be attributed for the foundation failures in tropical soils drawing on the Malaysia experience.

Lack of technology

A foundation may fail due to lack of technology, i.e. the state of the technology which is available is not adequate to cope with the problem at hand. Although this seems to be rarely the case, the situation does arise, and if it is due to ground difficulty, it is sometimes feasible to relocate the structure concerned to avoid the difficulty. If not, the client should be informed of the inherent risks which are being faced.

Inadequate site investigation

Some foundation failures are caused by inadequate site investigation. On rare occasions failures have been caused by the complete absence of site investigation. Both are due to ignorance, lack of expertise in site investigation or pressure from the client to save cost. It has been said that "you pay for a site investigation whether you have one or not" (Littlejohn 1991). This is because an inadequate site investigation can lead to either over-conservative design, which is wasteful and expensive, or to foundation difficulty or even failure.

A direct consequence of an inadequate site investigation is that it places both the design and construction processes at risk due to lack of vital ground information. In a 25-year study of civil engineering and building projects in the U.K., it has been found that the largest element of technical and financial risk lies normally in the ground (ICE 1991). On the other hand, an adequate site investigation is one in which the technical and financial risks of the foundation are kept within manageable proportions.

Geology should not be overlooked in site investigation practice. It is worthwhile to examine the geology of the site based on published geological information. Geology is a useful tool to assist in the planning of site investigation and interpretation of the results. This is because geology can give a broad indication of certain features which may be expected in a particular type of geology, e.g. boulders in granite areas. These features however need to be confirmed by a site investigation.

Design inadequacy

This is one of the two major causes of foundation failures. The essence of the foundation design for, say, a building is to provide a foundation, be it footing pile/s, at every column such that the column load is supported satisfactorily on the ground, i.e. the factor of safety with respect to load carrying capacity and the settlement of each foundation must be within certain prescribed limits. In addition, this has to be achieved with minimum cost. Unfortunately, the ground properties at every column position can vary significantly and sometimes by large amounts. Thus the foundation at every column must be individually designed with regard to size and depth to match the precise ground conditions at that location. It is not adequate to design for the average ground condition for the site. Indeed for certain types of pile foundation, e.g. large drilled (bored) piles in limestone areas, every pile within a pile cap must be individually designed.

Common shortcomings in design include wrong choice of foundation system or piling system, and inadequate depth or size of foundation. This may be because certain adverse ground conditions were misjudged. Sometimes the design shortcoming are due to the inability to account for all the forces arising from deformation, effects of time or lowering of ground water level; as, for example down-drag on piles, or horizontal loads on piles due to surcharge on the ground nearby. Vaughan (1994), in examining assumptions, predictions and reality in geotechnical engineering has concluded that "in dealing with engineering problems, the understanding of real behavior is more important than accurate calculation".

Construction inadequacy

This is the second major cause of foundation failures (Chan 1996). Common construction shortcomings in foundation include:

a. Setting out error of footings or piles;
b. Insufficient penetration of driven or drilled (bored) piles;
c. Defective concrete in footings or piles;
d. Damage to pile shaft due to driving;
e. Hole collapse in drilled (bored) pile causing necking, etc.;
f. Deflection of driven piles, e.g. in limestone areas; and
g. Bending of driven piles, especially slender steel piles.

One inherent difficulty associated with construction of deep foundations, e.g. drilled (bored) piles, arises from carrying out the operation below ground out of sight; and when completed the finished product is also not accessible. However, with modern advances in technology it is now feasible to conduct non-destructive testing on piles to assess integrity of pile shaft and dynamic load capacity using stress wave-methods, (ICE 1988). For driven concrete and steel piles, hollow tubes cast in or welded to the pile shaft permit checking of pile vertically and verification of pile length. This can be a deterrent against dishonest practices.

Static load testing remains the most reliable technique for assessing single pile performance. Such testing should be started early in the construction program to verify that both the design and construction are satisfactory.

Lack of design input during construction

The supervision of geotechnical work, especially foundations, require special consideration since this has a crucial impact on the success or failure of foundation construction.

Some foundation failures have occurred because many engineers and decision makers do not realize that design input is crucial in the construction stage. This is a common misconception. Unlike superstructures, "in applied soil mechanics a design is not completed until the construction is successfully completed", Casagrande (1965).

This unusual situation which is peculiar only to geotechnical engineering, has arisen because of the inability to detect in advance during the design stage all the possible significant properties and conditions of the ground. It is neither practical nor desirable, in terms of cost and time, to investigate a site to an extent which will attain the above mentioned level of ground knowledge. Therefore the designer will have to make some assumptions concerning the ground which may differ from reality. In view of the above, every foundation design is, to some extent, tentative and incomplete at the design stage. Indeed, it may be useful to insert a statement to this effect on every design/construction drawing.

The design is expected to be completed in the construction stage when more information is uncovered concerning the ground conditions. Thus, during construction, the designer is required to refine or even amend his foundation design to suit the real ground conditions. This is the basis of the well-known Observational Method (Peck 1969) which is widely practised in geotechnical engineering.

In order to do this successfully, steps will have to be taken during construction to gather the missing information usually through field observations which are made during excavation, boring, concreting, pile driving, etc. In this way, the unknown ground in between site investigation boreholes can be covered and properly assessed. In the construction stage, ground conditions must be never assumed, e.g. by drawing straight lines in between adjacent boreholes. Occasionally, further site investigation may also be carried out during construction.

Undiserable trends and practices

At present, there are certain undesirable trends and practices which tend to increase the prospects of foundation failures. These are largely due to either ignorance or commercial pressure.

In Malaysia, the foundation practice is governed by the Uniform Building Bye-Laws which are deemed to be satisfied if the relevant code of practice is complied with. The required factors of safety are clearly spelt out in the code. In recent years, owing to increasing competition there is a tendency to adopt structural factors of safety in piles which are too low. This is manifested in unrealistically high permissible stresses in steel and concrete piles, the latter including both the driven and bored piles. Occasionally, the structural design of piles is confused with the structural design of columns in the superstructure. It is worth noting that the factor of safety of a pile should reflect all the uncertainties associated with underground construction. Therefore the structural factor of safety for piles should be significantly higher than that for structural columns.

The risks of high permissible stresses in drilled (bored) piles have been highlighted by Ting and Chan (1991). If a higher characteristic concrete strength is specified, it does not mean that the permissible stress can be increased proportionately, since the concrete in bored piled is invariably not compacted, apart from contamination of bentonite, etc. One disadvantage of using high-strength concrete in bored piles is the increased risk of shrinkage cracking of the shaft (ICE 1988).

Occasionally, lump sum contract is adopted for foundation construction which should be avoided whenever possible (Chan 1996). The usual reason is that the client wants to fix the limit of his financial commitment and pass the financial risk of variation to the contractor. However, such a step has wide implications which are often not appreciated by clients. This is because at the tender stage, it is usually very difficult, sometimes impossible, to quantify all the technical and financial risks in the foundation construction. For example, in a difficult ground the real cost of foundation construction can exceed the estimated cost by 50% or more. As a result the tender has an element of "gambling" involved. The problem will arise if the ground conditions turn out to be far more unfavorable than were expected and this may place unbearable pressure on the contractor. In a worst case, the contractor may be tempted to choose between financial ruin or sacrificing the safety of a foundation, thereby jeopardizing the entire project. In view of the above, lump sum contract represent unsound business practice for both the contractor and the client.

REFERENCES

Bjerrum, L. 1963. Discussion on *European Conference on Soil mechanics and Foundation Engineering*, Wiesbaden: (2): 135.

Burland, J.B. 2004. Interaction between structural and geotechnical engineer. New perspectives in the design and construction of foundation in structures. Joint structural division annual seminar. Hong Kong. 1–29.

Bruce, D.A., DiMillio, A.F. & Juran, I. 1995. A primer on micropiles. *ASCE, Civil Engineering*, December. 51–54.

Casagrande, A. 1965. Role of the "calculated risk" in earthwork and foundation engineering. Terzaghi lecture. *ASCE Journal of Soil Mech. & Found*, Div. 91(SM4): 72–111.

Chan, S.F. 1996. Foundation failures. Sixth Professor Chin Fung Kee. Memorial lecture, Institution of Engineers Malaysia. Petaling Jaya: 1–20.

Chan, S.F. 2004. Foundation failures – a review. *Proceedings Malaysian Geotechnical Conference*. Institution of Engineers Malaysia. Subang: 335–356.

FPS. 1987. Specification for the construction of mini piles. Federation of piling specialists, UK.

Greenwood, D. 1987. Underpinning by grouting. *Ground Engineering Journal*. April 20(3): 21–30.

Gue, S.S. & Tan, Y.C. 2003. Prevention of failures of bridge foundation and approach embankment on soft ground. *Proceedings of 2nd International Conference on Advances in Soft Soil Engineering and Technology*. Putrajaya, Malaysia: 245–258.

Hong Kong GEO. 1994. Guide to pile design and construction. Geotechnical engineering office, Civil Engineering Dept., Hong Kong: 32–33.

Institution of Civil Engineers UK. 1988. *Specification for piling*. London: Thomas Telford.

Institution of Civil Engineers UK. 1991. *Report on "Inadequate site investigation"*. London: Thomas Telford.

Littlejohn, G. 1991. Foreword to report on "Inadequate site investigation", Institution of Civil Engineers UK. London: Thomas Telford.

Lizzi, F. 1982. The "pali radice" (root piles). *Symposium on Soil and Rock Improvement Techniques*, AIT. D-11 to 21.

Peck, R.B. 1969. Advantages and limitations of the observational method in applied soil mechanics, Rankine lecture. *Geotechnique*. 19(2): 171–187.

Peck, R.B. 1981. Where has all the judgement gone? Norwegian Geotechnical Institute, Oslo Publication (134): 1–5.

Peck, R.B. 1988. Foreword to "geotechnical instrumentation for monitoring field performance" by J. Dunnicliff and G.E. Green. John Wiley & Sons.

Poulos, H.G. & Davis, E.H. 1980. *Pile foundation analysis and design*. Canada: John Wiley and Sons, Inc.

Stocker, M.F. 1994. 40 years of micropiling, 20 years of soil nailing, where do we stand today? *Proc. XIII ICSMFE*. (5): 167–168.

Terzaghi, K. 1936. Relation between soil mechanics and foundation engineering. Presidential address. *Proceedings 1st Conference on Soil Mechanics and Foundation Engineering*. Cambridge, USA. 3: 13–18.

Tschebotarioff, G.P. 1973. *Foundation, retaining and earth structures*. McGraw-Hill Inc. USA.

The Star, 1980. Danger flats must go. 18th June issue.

The Star, 1993. Coming down – 16 new PKNS houses. 29 June issue.

Ting, W.H. 1979. Consolidation of partially saturated residual soil. *Proceedings 6th Asian Regional Conference on Soil Mechanics and Foundation Engineering*. Singapore. 1: 95–98.

Ting, W.H. & Chan, S.F. 1991. Soil-structure interaction and foundation. Theme lecture. *Proceedings 9th Asian Regional Conference on Soil Mechanics and Foundation Engineering*. Bangkok. 147–161.

Vaughan, P.R. 1994. Assumption, prediction and reality in geotechnical engineering. Rankine lecture. *Geotechnique* 44(4): 573–609.

CHAPTER 13

Special Topic: Pile Supported Embankment

Bujang B.K. Huat
Department of Civil Engineering, University Putra Malaysia, Malaysia

Mageswaran Pavadai
R & A Geotechnics Sdn. Bhd., Kuala Lumpur, Malaysia

13.1   INTRODUCTION

In areas where the compressible soil layer is very deep (thick), especially where large settlements could not be accepted, or other methods of constructions could not be employed such as in the case of construction over deep tropical peat, civil engineering structures such as road embankments can then be supported on a group of piles. By carrying the structural forces to a competent layer, the problem of settlement of the structure can be largely avoided. However, this technique is generally accepted as expensive in terms of initial construction cost, and is therefore only be used in special circumstances. A cost estimate made by Greenacres (1996) showed that the fabric (geotextile)-reinforced embankment was five times more expensive than normal earth embankment, and a piled embankment was five times more expensive than the former.

13.2   DESIGN OF PILE SUPPORTED EMBANKMENT

The design of a pile system for an embankment is similar in principle to the foundation design to support other structures. The pile foundation is needed under sections where settlements are to be kept to minimum, such as adjacent to a road culvert, overhead bridge or bridge for a river crossing. In these cases, the piles are usually driven to set.

Piles are usually driven into the ground arranged in a grid form in the plan area, and fitted with individual caps. Figure 13.1 shows an empirical relationship that can be used as a rough guide to determine size of the pile cap, a, and distance between pile, s. Overburden load in general is transferred to the pile through an arching mechanism in the earth fill, as illustrated in Figure 13.2. To ensure this mechanism occurs, and for long-term stability, earth fill, especially the lower layer must comprise high quality fill, such as rock fill or crushed stones (Huat & Ali 1993; Huat et al. 1994). Sometimes synthetic fabric (geogrids) may also be used for this purpose (Reid & Buchanan 1984).

It is also of interest to note that studies carried out by Huat (1991) show that in a piled embankment design, the fill thickness, depending on the area ratio of the pile cap (a^2/s^2), needs to be at least equal to, or greater than the pile spacing in order to prevent failure by punching. The pile support is very sensitive to the area ratio (a^2/s^2). Reducing the area ratio may lead to a higher load on an individual pile but reduces the value of efficacy, E and throws more loads onto the subsoil. Efficacy, E is defined as pile load over total available load, $\gamma H s^2$ (γ is the unit weight of fill, H is the height of fill above the pile head, and s is the pile spacing). The efficacy of the pile support can be increased by increasing the area ratio of the pile support, achieved by either enlarging the pile cap or reducing the pile spacing. Since the transfer of fill load from ground to the top of the pile cap relies on the shearing resistance of the fill, type or quality of fill will therefore affect the efficacy of the pile support. Better quality fills such as rock fill or crushed stones would facilitate more efficient transfer of fill load to the piles. Figure 13.3 shows comparison of pile support efficacy E with H/s ratio from various sources; closed form solutions, case histories and model studies.

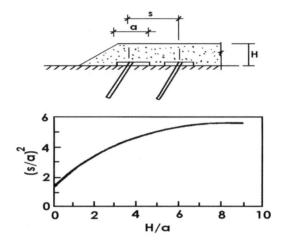

Figure 13.1.   Design chart of pile embankment (from Swedish Road Board 1974).

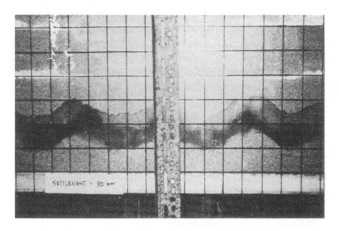

Figure 13.2.   Arching mechanism in the earth fill (Huat & Ali 1993).

It is of interest to note that there is no significant increase in rate of efficacy once the fill height exceeds about two times the pile spacing. It may therefore be advocated that a high quality fill (such as rock or crushed stone) need not be higher than about two times the pile spacing above the pile head. A lower grade fill may be placed above this elevation, the overburden of which will be transferred to the piles via arching of the lower layer. As mentioned before, a layer of relatively stiff fabric (or geogrids) placed immediately above the pile head may also help to improve the local bearing capacity of the pile support.

In cases where the final embankment height is low relative to the pile spacing, or the pile spacing is too large (too much load is available to be relieved), a near vertical rupture or punching occurs throughout the whole depth of fill above the pile head. This rupture mechanism is shown in Figure 13.4.

A closed form solution, propose by Hewlett & Randolph (1988) to estimate the pile efficacy E, is based on limiting the equilibrium of forces required to sustain an arch for free draining granular material base on the trap door theory (Figure 13.5), and given as follows:

For low fill,

$$E = \frac{s^2 - a^2}{s^2 \gamma H} A_p \tag{13.1}$$

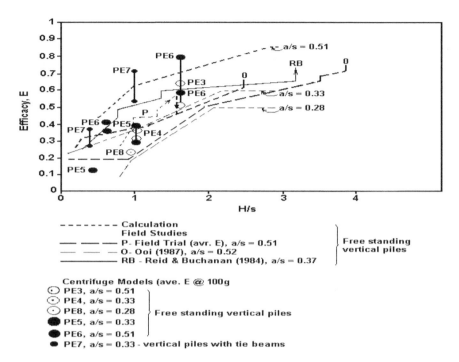

Figure 13.3.   Comparison of pile support efficacy, E (Huat & Ali 1993, Huat et al. 1994).

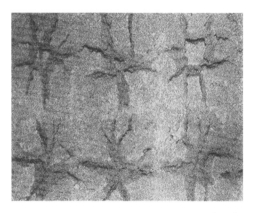

Figure 13.4.   Punching failure (view from top; for the case of $H/s = 1$, $a^2 = s^2 = 0.125$) (Huat et al. 1994).

where $A_p = B_p + \gamma(s - a)/\sqrt{2}$

$$B_p = \left[\gamma\left(1 - (C_p)^{2(k_p-1)}\right)\right]\left[H - \frac{s}{\sqrt{2}}\left(\frac{k_p - 1}{2k_p - 3}\right)\right] + \gamma\frac{s - a}{\sqrt{2}(2k_p - 3)}$$

$C_p = a/s$

k_p = passive earth pressure coefficient ($= tan^2(45 + \phi/2)$), and φ is the fill angle of friction.

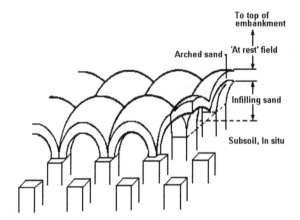

Figure 13.5.   Isometric view of a grid of pile caps and a series of domes forming a vault spanning between them (Hewlett & Randolph 1988).

s = pile spacing
γ = unit weight of fill
H = height of fill above the pile head
For high fill (or failure at pile cap)

$$E = \frac{D_p}{1 + D_p} \tag{13.2}$$

where $D_p = \dfrac{2k_p}{(k_p + 1)}\dfrac{1}{(1 + C_p)}\left[(1 - C_p)^{-k_p} - (1 + C_p)k_p\right]$

The lower value of the two estimates is used for design.

If the bearing capacity of the surface soil layer is not sufficient to allow for interaction with the piles, or if the embankment to be constructed is too high, concrete deck (or also known as piled raft) may be built on the ground surface on top of the piles. Concrete deck or piled raft supported on pile is expansive but will not be as expansive as a flyover structure. Alternatively geosynthetics such as geogrids can be laid on top of the pile cap. Figure 13.6 shows typical cross sections of the pile embankment, i.e. for the case of a pile with RC (reinforced concrete) slab as concrete deck, and piles with individual caps and geosynthetic fabric.

For pile embankment construction, usually driven piles are used. Lateral displacement and large deformation will occur in the subsoil especially if the soil is very sensitive. As this may cause problems of instability in nearby structures, precautions need to be taken. The technique of pile embankment has been successfully used in Malaysia especially for construction of high road embankments on soft soils such as alluvial and marine clays, and peat.

There is another one major point that needs very careful consideration with this method of construction; that is the relative movement between the supported and unsupported sections of the structure/embankment. In the past, many problems have been encountered due to this relative movement (or differential settlement). Failures had even resulted due to the large differentials settlement between the piled and unpiled section of an embankment, as shown in Figure 13.7.

There is a need to provide a good transition zone between the completely piled structures (such as the bridge or culvert), and the settling earth embankment. In this transition zone, the piles are driven to a variable depth (not to set), in order to grade as uniformly as possible the settlements. The lengths of each row of these so-called "floating" piles are steadily reduced over a distance to achieve the specified differential settlement tolerance. A typical section of the transition zone is shown in Figure 13.8.

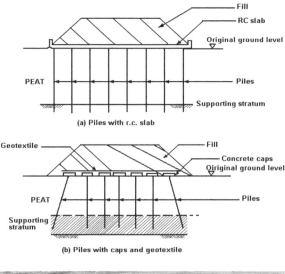

(a) Piles with r.c. slab

(b) Piles with caps and geotextile

(c) Grid of (timber) piles. The RC slab is shown in the
background.

Figure 13.6.   Typical cross sections of pile embankment.

Design of piled RC slab

An illustration is given of the design of a concrete deck or piled raft of a 2.5 m high embankment supported on 150 × 150 mm square RC piles. Figure 13.9 shows the embankment section.

Limitations/assumptions considered in the design:

– The slab is assumed to be supported on a regular grid of piles.
– The slab is assumed to be a one-way continuous slab analysed and designed separately in both x and y directions.
– The most likely mode of failure is assumed to be the folded plate mechanism.
– Internal spans are assumed to have the same moment at both ends.
– Top reinforcement over the internal supports is concentrated in a strip half the width of the span; this arrangement provides better performance with regard to punching shear. Top reinforcement over the external supports is concentrated in a strip one-fifth the width of the span. The reinforcement is designed by using an appropriate multiplier in the calculation of the support moment.
– Traffic load is considered as 20 kPa.
– The loading from the embankment fill and traffic is assumed to be transferred to the piles via the pile caps.

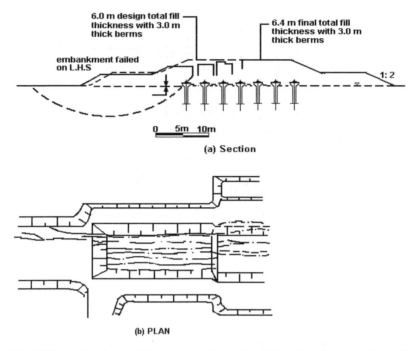

Figure 13.7.   Differential settlement failure between piled and unpiled section of an embankment (Huat et al. 1994).

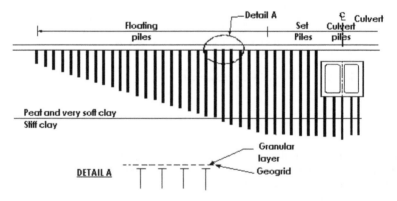

Figure 13.8.   Transition zone.

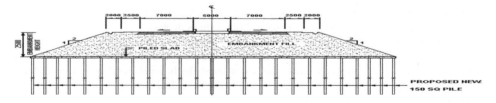

Figure 13.9.   Cross section of a 2.5 m high piled raft embankment.

Table 13.1.   Pile spacing tabulation.

Embankment height (m)	Traffic load distribution at cap level (kN/m²)	Total unfactored pressure at the piled raft level (kN/m²)	Pile distance, s in square pattern based on load schedule (m)
2.5	20	72.2	1.61

With due consideration of the above, the following listed parameters are taken into account:

- Max. allowable pile load   160 kN*
  (150 × 150 mm square precast RC pile)
- Mass of the fill material   18 kN/m²
- Internal friction of fill material   27°

* based on the recommendation of precast RC pile supplier.

Based on the above design parameters and design criterion, we shall proceed with embankment design calculation which consists of the following major steps.

Step 1: Computation of pile spacing
The pile spacing, s, is computed based on the formula below:

$$s = \sqrt{\frac{F_p}{(f_{fs} \times H \times \gamma) + (f_f \times w_s) + (f_q \times q_s)}}$$

where F_p = pile working load (kN); H = embankment height (m); γ = unit weight of fill (kN/m³); w_s = traffic load (kN/m²); q_s = dead load (kN/m²).
Load factors as per BS8006 Table 27 for Ultimate Limit State Conditions are:

Load factor of embankment fill: $f_{fs} = 1.3$
Load factor of dead loads: $f_f = 1.2$
Load factor of live loads: $f_q = 1.3$

Based on this formula, the pile spacing, s, in square pattern is calculated for the center of the embankment with an embankment height of 2.5 m high, with (assumed) piled slab thickness of 300 mm and traffic load of 20 kPa. The results of the analysis are tabulated in Table 13.1 above.

Step 2: Design of piled raft
The piled raft is designed considering it as a flat slab. A detailed computation of the slab reinforcement together with punching check is shown below:

Design of flat slab in accordance to BS8110: Part 1:1997

Slab geometry
Span of slab in x-direction/pile spacing x-direction;   Span$_x$ = **1610** mm
Span of slab in y-direction/pile spacing x-direction;   Span$_y$ = **1610** mm
Column dimension in x-direction; l_x = **150** mm
Column dimension in y-direction; l_y = **150** mm
External column dimension in x-direction; l_{x1} = **150** mm
External column dimension in y-direction; l_{y1} = **150** mm
Edge dimension in x-direction; $e_x = l_{x1}/2 = $ **75** mm
Edge dimension in y-direction; $e_y = l_{y1}/2 = $ **75** mm
Effective span of internal bay in x direction; $L_x = $ Span$_x - l_x = $ **1460** mm
Effective span of internal bay in y direction; $L_y = $ Span$_y - l_y = $ **1460** mm
Effective span of end bay in x direction; $L_{x1} = $ Span$_x - l_x/2 = $ **1535** mm
Effective span of end bay in y direction; $L_{y1} = $ Span$_y - l_y/2 = $ **1535** mm

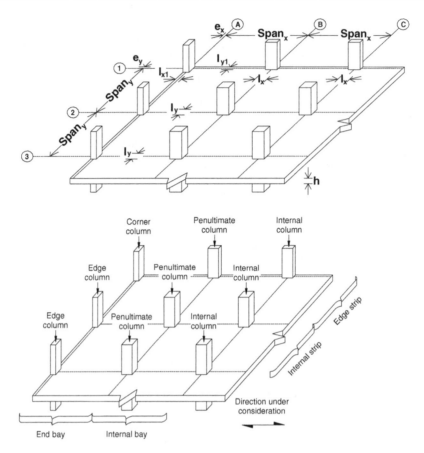

Figure 13.10.   Illustration to typical notations used in the calculation.

Slab details

Depth of slab;	$h = 300$ mm
Characteristic strength of concrete;	$f_{cu} = 35$ N/mm^2
Characteristic strength of reinforcement;	$f_y = 460$ N/mm^2
Characteristic strength of shear reinforcement;	$f_{yv} = 460$ N/mm^2
Material safety factor;	$\gamma_m = 1.05$
Cover to bottom reinforcement;	$c = 75$ mm
Cover to top reinforcement;	$c' = 35$ mm

Loading details

Characteristic dead load;	$G_k = 52.200$ kN/m^2
Characteristic imposed load;	$Q_k = 20.000$ kN/m^2
Dead load factor;	$\gamma_G = 1.4$
Imposed load factor;	$\gamma_Q = 1.6$
Total ultimate load;	$N_{ult} = (G_k \times \gamma_G) + (Q_k \times \gamma_Q)$
	$= 105.080$ kN/m^2
Moment redistribution ratio;	$\beta_b = 1.1$
Ratio of midspan moments to support moments;	$i = 1.0$

Design of slab in the x-direction

The slab is checked only in one direction as the spacing in both directions is identical. Figure 13.10 shows an illustration of typical notations used in the calculation.

SAGGING MOMENTS

End bay A-B

Effective span;	$L = \mathbf{1535}\,\text{mm}$
Depth of reinforcement;	$d = \mathbf{207}\,\text{mm}$
Midspan moment;	$m = (N_{ult} \times L^2)/(2 \times (1 + \sqrt{(1+i)})^2) = \mathbf{21.240}\,\text{kNm/m}$
Support moment;	$m' = i \times m = \mathbf{21.240}\,\text{kNm/m}$

Design reinforcement
Lever arm;
$$K' = 0.402 \times (\beta_b - 0.4) - 0.18 \times (\beta_b - 0.4)^2 = \mathbf{0.193}$$
$$K = m/(d^2 \times f_{cu}) = \mathbf{0.014}$$
Compression reinforcement is not required
$$z = \min(0.5 + \sqrt{(0.25 - (K/0.9))}, 0.95) \times d = \mathbf{196.6}\,\text{mm}$$
Area of reinforcement required;
$$A_{s\_req} = m/(z \times f_y/\gamma_m) = \mathbf{247}\,\text{mm}^2/\text{m}$$

Provide 12 dia bars @ 300 centres

Area of reinforcement provided;
$$A_{s\_prov} = \pi \times D^2/(4 \times s) = \mathbf{377}\,\text{mm}^2/\text{m}$$

PASS – Span reinforcement is OK

Check deflection

Design service stress;	$f_s = 2 \times f_y \times A_{s\_req}/(3 \times A_{s\_prov} \times \beta_b) = \mathbf{182}\,\text{N/mm}^2$
Modification factor;	$k_1 = 0.55 + (477\,\text{N/mm}^2 - f_s)/(120 \times (0.9\,\text{N/mm}^2 + (m/d^2)))$ $= \mathbf{2.309}$
Allowable span to depth ratio;	$0.9 \times 26 \times k_1 = \mathbf{54.041}$
Actual span to depth ratio;	$L/d = \mathbf{7.415}$

PASS – Span to depth ratio is OK

Internal bay B-C

Effective span;	$L = \mathbf{1460}\,\text{mm}$
Depth of reinforcement;	$d = \mathbf{207}\,\text{mm}$
Midspan moment;	$m = (N_{ult} \times L^2)/(2 \times (\sqrt{(1+i)} + \sqrt{(1+i)})^2) = \mathbf{13.999}\,\text{kNm/m}$
Support moment;	$m' = i \times m = \mathbf{13.999}\,\text{kNm/m}$

Design reinforcement
Lever arm;
$$K' = 0.402 \times (\beta_b - 0.4) - 0.18 \times (\beta_b - 0.4)^2 = \mathbf{0.193}$$
$$K = m/(d^2 \times f_{cu}) = \mathbf{0.009}$$
Compression reinforcement is not required
$$z = \min(0.5 + \sqrt{(0.25 - (K/0.9))}, 0.95) \times d = \mathbf{196.6}\,\text{mm}$$
Area of reinforcement required;
$A_{s\_req} = m/(z \times f_y/\gamma_m) = \mathbf{162}\,\text{mm}^2/\text{m}$
Provide 12 dia bars @ 300 centres
Area of reinforcement provided;
$A_{s\_prov} = \pi \times D^2/(4 \times s) = \mathbf{377}\,\text{mm}^2/\text{m}$
PASS – Span reinforcement is OK

Check deflection

Design service stress;	$f_s = 2 \times f_y \times A_{s\_req}/(3 \times A_{s\_prov} \times \beta_b) = \mathbf{120}\,\text{N/mm}^2$
Modification factor;	$k_1 = 0.55 + (477\,\text{N/mm}^2 - f_s)/(120 \times (0.9\,\text{N/mm}^2 + (m/d^2)))$ $= \mathbf{2.974}$
Allowable span to depth ratio;	$0.9 \times 26 \times k_1 = \mathbf{69.593}$
Actual span to depth ratio;	$L/d = \mathbf{7.053}$

PASS – Span to depth ratio is OK

HOGGING MOMENTS – INTERNAL STRIP

Penultimate column B3

Consider the reinforcement concentrated in half width strip over the support

Depth of reinforcement; $d' = 247$ mm

Support moment; $m' = 2 \times i \times m = 42.480$ kNm/m

Lever arm; $K' = 0.402 \times (\beta_b - 0.4) - 0.18 \times (\beta_b - 0.4)^2 = 0.193$

$$K = m'/(d'^2 \times f_{cu}) = 0.020$$

Compression reinforcement is not required

$$z = \min((0.5 + \sqrt{(0.25 - (K/0.9))}), 0.95) \times d' = 234.7 \text{ mm}$$

Area of reinforcement required; $A_{s\_req} = m'/(z \times f_y/\gamma_m) = 413 \text{ mm}^2/\text{m}$

Provide 12 dia bars @ 250 centres

Area of reinforcement provided; $A_{s\_prov} = \pi \times D^2/(4 \times s) = 452 \text{ mm}^2/\text{m}$

PASS – Support reinforcement is OK

Internal column C3

Consider the reinforcement concentrated in half width strip over the support

Depth of reinforcement; $d' = 247$ mm

Support moment; $m' = 2 \times i \times m = 27.999$ kNm/m

Lever arm; $K' = 0.402 \times (\beta_b - 0.4) - 0.18 \times (\beta_b - 0.4)^2 = 0.193$

$$K = m'/(d'^2 \times f_{cu}) = 0.013$$

Compression reinforcement is not required

$$z = \min((0.5 + \sqrt{(0.25 - (K/0.9))}), 0.95) \times d' = 234.7 \text{ mm}$$

Area of reinforcement required; $A_{s\_req} = m'/(z \times f_y/\gamma_m) = 272 \text{ mm}^2/\text{m}$

Provide 12 dia bars @ 300 centres

Area of reinforcement provided; $A_{s\_prov} = \pi \times D^2/(4 \times s) = 377 \text{ mm}^2/\text{m}$

PASS – Support reinforcement is OK

Based on the same methodology as above, the hogging reinforcement check is done for the external strip.

Punching shear check

Only penultimate edge column and penultimate central column are checked for punching as they suffer the most load.

Penultimate edge column

Design shear transferred to column; $V_t = ((0.45 \times \text{Span}_x) + e_x) \times (1.05 \times \text{Span}_y)$
$$\times N_{ult} = 142 \text{ kN}$$

Design effective shear transferred to column; $V_{eff} = 1.25 \times V_t = 178$ kN

Area of tension steel in x-direction; $A_{sx\_ten} = A_{sx\_edge} = 376 \text{ mm}^2/\text{m}$

Area of tension steel in y-direction; $A_{sy\_ten} = A_{sy1e} = 411 \text{ mm}^2/\text{m}$

Column perimeter; $u_c = (2 \times l_{x1}) + l_y = 450$ mm

Average effective depth of reinforcement; $d = h - c - \phi_p = 213$ mm

Maximum allowable shear stress; $v_{max} = \min(0.8 \times \sqrt{(f_{cu})}, 5) = 4.733 \text{ N/mm}^2$

Design shear stress at column perimeter; $v_0 = V_{eff}/(u_c \times d) = 1.852 \text{ N/mm}^2$

PASS – Maximum concrete shear stress not exceeded at column perimeter

Shear reinforcement at a perimeter of 1.50d – (319 mm)

Length of shear perimeter; $u = u_c + (2 \times (k_x \times k_y) \times k \times d) = 1728$ mm

Area of tension steel at shear perimeter; $A_{s\_ten} = (k_y \times (p_x + (k_x \times k \times d)) \times A_{sy\_ten})$
$$+ (k_x \times (p_y + (k_y \times k \times d)) \times A_{sx\_ten})$$
$$A_{s\_ten} = 683 \text{ mm}^2$$

Design concrete shear stress;

$$v_c = 0.79 \times \min(100 \times A_{s\_ten}/(u \times d), 3)^{1/3}$$
$$\times \max(400/d, 1)^{1/4}/1.25$$
$$v_c = \mathbf{0.422} \text{ N/mm}^2$$

Nominal design shear stress at perimeter;

$$v = V_{eff}/(u \times d) = \mathbf{0.482} \text{ N/mm}^2$$
$$\mathbf{v_c < v < = 1.6 \times v_c}$$

Shear reinforcement required at perimeter;

$$A_{sv\_req} = (v - v_c) \times u \times d/(0.95 \times f_{yv}) = \mathbf{51} \text{ mm}^2$$

Shear reinforcement at a perimeter of 2.25d – (479 mm)

Length of shear perimeter;

$$u = u_c + (2 \times (k_x \times k_y) \times k \times d) = \mathbf{2367} \text{ mm}$$

Area of tension steel at shear perimeter;

$$A_{s\_ten} = (k_y \times (p_x + (k_x \times k \times d)) \times A_{sy\_ten})$$
$$+ (k_x \times (p_y + (k_y \times k \times d)) \times A_{sx\_ten})$$
$$A_{s\_ten} = \mathbf{934} \text{ mm}^2$$

Design concrete shear stress;

$$v_c = 0.79 \times \min(100 \times A_{s\_ten}/(u \times d), 3)^{1/3}$$
$$\times \max(400/d, 1)^{1/4}/1.25$$
$$v_c = \mathbf{0.422} \text{ N/mm}^2$$

Nominal design shear stress at perimeter;

$$v = V_{eff}/(u \times d) = \mathbf{0.352} \text{ N/mm}^2$$
$$\mathbf{v < v_c \text{ no shear reinforcement required}}$$

Penultimate central column

Design shear transferred to column;

$$V_t = (1.05 \times \text{Span}_x) \times (1.05 \times \text{Span}_y)$$
$$\times N_{ult} = \mathbf{300} \text{ kN}$$

Design effective shear transferred to column;

$$V_{eff} = 1.15 \times V_t = \mathbf{345} \text{ kN}$$

Area of tension steel in x-direction;

$$A_{sx\_ten} = A_{sx1e} = \mathbf{452} \text{ mm}^2/\text{m}$$

Area of tension steel in y-direction;

$$A_{sy\_ten} = A_{sy1e} = \mathbf{411} \text{ mm}^2/\text{m}$$

Column perimeter;

$$u_c = 2 \times (l_x + l_y) = \mathbf{600} \text{ mm}$$

Average effective depth of reinforcement;

$$d = h - c - \phi_p = \mathbf{213} \text{ mm}$$

Maximum allowable shear stress;

$$v_{max} = \min(0.8 \times \sqrt{(f_{cu})}, 5) = \mathbf{4.733} \text{ N/mm}^2$$

Design shear stress at column perimeter;

$$v_0 = V_{eff}/(u_c \times d) = \mathbf{2.702} \text{ N/mm}^2$$

PASS – Maximum concrete shear stress not exceeded at column perimeter

Shear reinforcement at a perimeter of 1.50d – (319 mm)

Length of shear perimeter;

$$u = u_c + (2 \times (k_x \times k_y) \times k \times d) = \mathbf{3156} \text{ mm}$$

Area of tension steel at shear perimeter;

$$A_{s\_ten} = (k_y \times (p_x + (k_x \times k \times d)) \times A_{sy\_ten})$$
$$+ (k_x \times (p_y + (k_y \times k \times d)) \times A_{sx\_ten})$$
$$A_{s\_ten} = \mathbf{1362} \text{ mm}^2$$

Design concrete shear stress;

$$v_c = 0.79 \times \min(100 \times A_{s\_ten}/(u \times d), 3)^{1/3}$$
$$\times \max(400/d, 1)^{1/4}/1.25$$
$$v_c = \mathbf{0.435} \text{ N/mm}^2$$

Nominal design shear stress at perimeter;

$$v = V_{eff}/(u \times d) = \mathbf{0.514} \text{ N/mm}^2$$
$$\mathbf{v_c < v < = 1.6 \times v_c}$$

Shear reinforcement required at perimeter;

$$A_{sv\_req} = (v - v_c) \times u \times d/(0.95 \times f_{yv}) = \mathbf{122} \text{ mm}^2$$

Shear reinforcement at a perimeter of 2.25d – (479 mm)

Length of shear perimeter;

$$u = u_c + (2 \times (k_x \times k_y) \times k \times d) = \mathbf{4434} \text{ mm}$$

Area of tension steel at shear perimeter;

$$A_{s\_ten} = (k_y \times (p_x + (k_x \times k \times d)) \times A_{sy\_ten})$$
$$+ (k_x \times (p_y + (k_y \times k \times d)) \times A_{sx\_ten})$$
$$A_{s\_ten} = \mathbf{1913} \text{ mm}^2$$

Design concrete shear stress;

$$v_c = 0.79 \times \min(100 \times A_{s\_ten}/(u \times d), 3)^{1/3}$$
$$\times \max(400/d, 1)^{1/4}/1.25$$
$$v_c = \mathbf{0.435} \text{ N/mm}^2$$

Nominal design shear stress at perimeter;

$$v = V_{eff}/(u \times d) = \mathbf{0.366} \text{ N/mm}^2$$
$$\mathbf{v < v_c \text{ no shear reinforcement required}}$$

Based on the above design, the design reinforcement is summarized in Figure 3.11.

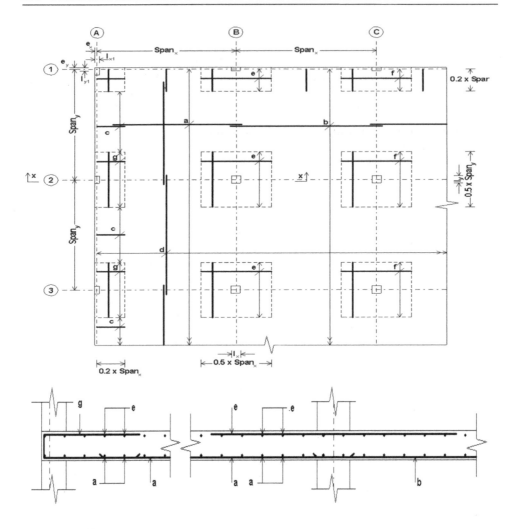

Section x-x

a = 12 dia bars @ 300 centres - (376 mm²/m);   b = 12 dia bars @ 300 centres - (376 mm²/m)
c = 12 dia bars @ 300 centres - (376 mm²/m);   d = 12 dia bars @ 300 centres - (376 mm²/m)
e = 12 dia bars @ 250 centres - (452 mm²/m);   f = 12 dia bars @ 300 centres - (376 mm²/m)
g = 12 dia bars @ 300 centres - (376 mm²/m)

Figure 13.11.   Arrangement of reinforcement in the slab.

REFERENCES

British Standards BS 8006: 1995. Section 8. Design of embankments with reinforced soil foundations on poor ground. British Standard Institution, UK.
British Standards BS 8110: 1997 Section 8. Design of concrete structures. British Standard Institution, UK.
Greenacres, M. 1996. Banking on soft options. *Ground Engineering*. 34–36.

Hewlett, J.M. & Randolph, M.F. 1988. Analysis of piled embankment. *Ground Engineering* 21(3): 12–18.

Huat, B.B.K. 1991. Simulation of field trial structures. PhD Thesis. Univ. of Manchester, U.K. (unpublished).

Huat, B.B.K. & Ali, F.H. 1993. A contribution to the design of a piled embankment, *Pertanika Journal of Science and Technology.* 1(1): 79–92.

Huat, B.B.K., Craig, W.H. & Ali, F.H. 1994. The mechanics of piled embankment. *FHWA International Conference on Design and Construction of Deep Foundations.* Orlando. Florida, USA. 2: 1069–1082.

Reid, W.M. & Buchanan, N.W. 1984. Bridge approach support piling. In *piling and ground treatment.* London: Telford. 267–274.

Swedish Road Board, 1974. Embankment piles. Report TV 121.

CHAPTER 14

Country Case Study: Engineering Geology in Relation to Foundation and Rock Slope Engineering in Malaysia

B.K. Tan

Faculty of Science & Technology, University Kebangsaan Malaysia, Bangi, Malaysia

14.1 INTRODUCTION

Engineering geology is an applied science dealing with the application of geology and geological methods in civil and construction works. The importance of geology as applied to the development of cities and general civil engineering works has been emphasised repeatedly by Legget (1973), Legget & Karrow (1983), Tan (1991), and many others. Numerous case studies can be found in the literature on the application of engineering geology in various engineering projects; perhaps the handbook by Legget & Karrow (1983) would be a good starting point for those interested in reading more about the use of geology in civil engineering works worldwide.

In Malaysia, the use of engineering geologists in civil engineering works has been on the rise in the past two decades or so, in particular during the "construction boom" periods. Mega-projects such as the North-South Expressway, the Light Rail Transit, the Kuala Lumpur City Centre (KLCC) Petronas Twin Towers, the Bakun Dam, etc. plus numerous other construction projects had witnessed some significant input by engineering geologists, or at least where the geological factors came into play and had to be contended with by the project engineers. Undoubtedly, this role of the geologist in civil construction works will be a permanent feature for all future projects.

This chapter discusses the roles or applications of engineering geology in foundation and rock slope engineering, two major activities in civil engineering works. Fundamentals of engineering geology are first discussed, followed by applications to foundations and rock slopes, and then some local case studies as illustrations. The topics on engineering geology and its applications to various civil engineering works, including foundation and rock slope engineering, have been discussed at various times by the author previously. Hence, materials in this chapter are based mainly on numerous previous publications by the author, notably: Tan (1982, 1991, 1999a, 2004a, 2004b and 2004c), among others. Several case histories are presented here.

These case histories cover different types of engineering works or projects involving foundations or rock slopes. They are also based mainly on the author's personal experiences or involvement with civil and geotechnical engineers in the last two decades or so. Many of these case studies have also been presented/published elsewhere, so this chapter provides a summary or compilation of some of these case histories. Further details of some of these case histories or projects can be referred to in the references cited.

14.2 FUNDAMENTALS OF ENGINEERING GEOLOGY

Engineering geology encompasses three fundamental studies or issues, namely: the lithology or rock type, geological structures, and weathering grades. These three fundamental issues are always incorporated in any engineering geologic studies, Tan (1998a). Together, they determine the mass characteristics and engineering behaviour of the rock formations encountered at a project site.

Lithology

Lithology means rock type. However, since superficial soils are also important in construction works, lithology here would include rock and soil types, or material types. Lithology is the first consideration since different rock types (and similarly soil types) would have different properties and behaviour. For example, granitic rocks differ from shales or schists. Limestone is yet another type of rock with its unique properties. Owing to the different nature and origin of the rock types mentioned above, the inherent geological structures associated with each rock type are also different. Granitic rocks are often intersected by three or more sets of joints as major structures, shales are dominated by bedding planes, schists are foliated, while limestones are characterised by their unique solution or karstic features. As such, for example, a rock slope cut into these different rocks would have to contend with different geological features or problems.

In addition, the residual soils derived from the weathering of the different rock types mentioned above would differ widely in soil properties. For example, granitic soils are usually sandy, while

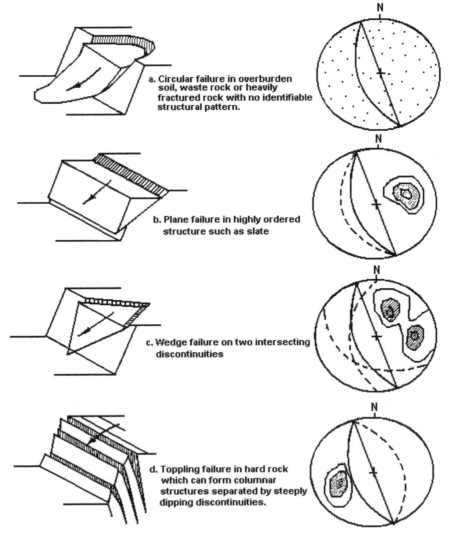

Figure 14.1.   Main types of slope failures and appearance of stereoplots of structural conditions likely to give rise to these failures (Hoek & Bray 1974).

shales produce silty to clayey soils. Insoluble residues accumulated after the solution of the limestone form unique residual red clays called *terra rossa*.

Structures

Geological structures include a host of features such as joints, faults, bedding planes, foliations, dykes, folds, etc. In general, the major fracture planes such as faults and major joints are the more critical structures since they represent breaks or weaknesses in the rock mass. Rock slope stability, for example, is controlled by the fracture planes. The engineering of rock slopes is based on detailed measurements and analysis of discontinuities using the stereonets, Figure 14.1, Hoek & Bray (1974), Tan (1999b). Dykes and sills are yet other geological structures that can be important in engineering works. In a recent major cut-slope failure, for example, the occurrence of an aplite dyke played a major role in the failure, Ting & Nithiaraj (1998).

Similarly, the fracture systems in the underlying rock mass would determine seepage losses in dam foundations, bearing capacity, and settlement of buildings. The fracture system, weathering grade, also determine the rock mass quality, Q, intersected by a tunnel, for purpose of determining tunnel stability and support measures as well as seepage control, Barton et al. (1974), Kveldsvik & Karlsrud (1995).

Weathering

Weathering is especially important in humid, tropical regions like Malaysia, since intense chemical weathering reduces rocks to soil-like materials, resulting in thick soil mantles over bedrock formations. Again, the details and thickness of soil cover over rock in weathering profiles differ from rock to rock. In granitic rocks, for example, the thickness of the soil zone is generally around 30 m, reaching up to 50 m in some extreme cases. In sedimentary (e.g. sandstone, shale) and metasedimentary (e.g. quartzite, phyllite, schist) rocks, the soil zone is thinner, say around 10 m only. For engineering purposes, the standard six-grade weathering classification as proposed by Little (1969), ISRM (1977), is most useful and has been used frequently by many workers, Figure 14.2. Some local experiences on weathering of rocks in relation to engineering have been summarised recently by the author, Tan (1995a).

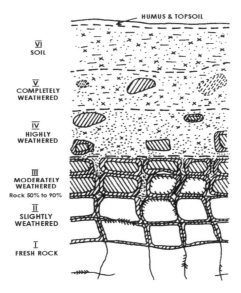

Figure 14.2.   Classification of residual tropical soils by degree of weathering (Little 1969).

Figure 14.3(a).   Limestone bedrock with massive overhang/sinkhole.

Superimposed on the lithology and structures, the weathering effects can be predominant in controlling the behaviour of the resultant rock or soil materials. For example, in a granitic weathering profile, the grades IV to VI zones behave as soil, while the grades I to III zones behave as rock (rock mechanics applies). This weathering effect is most pronounced and clearly illustrated in the treatment of cut slopes in hilly terrain – in analysis, design, excavation, stabilization, etc. of the cut slopes; utilizing either soil engineering or rock mechanics methods depending on the grades of weathering.

Similarly for a construction site, the nature or consistencies of the sub-surface soil strata would depend on the weathering profiles. In residual soils, for example, the soil strata would generally show increasing stiffness with depth, as evidenced by increasing Standard Penetration Test values with depth. Hence the deeper the soil stratum, the higher the bearing capacity and shear strength.

14.3   ENGINEERING GEOLOGY IN RELATION TO FOUNDATION ENGINEERING

Foundations in limestone

Limestone is rather unique since it is soluble in even slightly acidic waters to produce a host of solution or karstic features, with its associated engineering problems, now well known to many civil and geotechnical engineers. Many papers dealing with foundations in limestone are contained in two major publications by IEM, namely the Proceedings of the 8th and 12th S.E. Asian Geotechnical Conference held in Kuala Lumpur in 1985 and 1996 respectively. In addition, a recent Proceedings of the IEM-GSM Forum on "Karst: Geology & Engineering" (1999), also focus on the geology and engineering of limestone.

In the site investigation of limestone bedrock for foundation purposes, the adequacy of the number of boreholes drilled to sufficient depths to define the intricate nature of the limestone bedrock, solution channels, cavities, slumped zones, etc. cannot be over-emphasised. The number of boreholes for a highrise building site underlain by limestone bedrock now generally reaches over 100, with the KLCC Petronas Twin Towers project in Kuala Lumpur setting the record at ~400, Tarique Azam (1996). Incidentally, the shifting of the Twin Towers by about 50 m from their original proposed positions was partly related to the geologic setting at the site (limestone versus Kenny Hill formation, etc.). Figure 14.3 shows some limestone bedrock profiles for several high-rise building sites in Kuala Lumpur. Besides pinnacled bedrock profiles, sinkholes, overhangs, etc., the contact zones between the limestone and other rock formations such as granite and the Kenny Hill

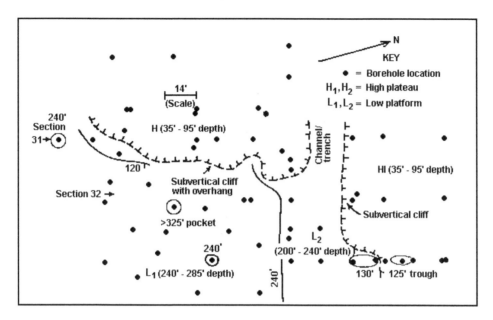

Figure 14.3(b).   Plan showing limestone bedrock at two levels and linear trenches, troughs and overhangs. Pan Pacific Hotel, Kuala Lumpur Malaysia.

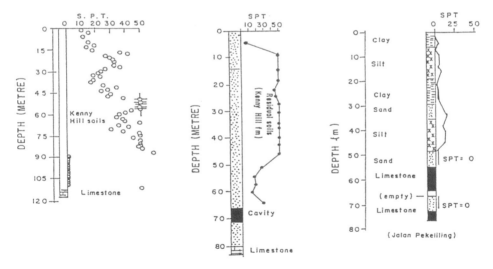

Figure 14.4.   Weak collapsed soil zone (slumped zone) with SPT = 0 above limestone bedrock, Central Kuala Lumpur, Malaysia.

formation are of great interest to engineers and geologists alike. The contact zones often exhibit deep solution channels or troughs due to differential and preferential erosion/solution along the limestone side of the contact zone.

In addition to the irregular or highly pinnacled bedrock profile and solution cavities, the occurrence of very weak slumped materials immediately above the limestone bedrock is yet another geological feature of great concern to the foundation engineers. These weak slumped materials are characterised by their very low Standard Penetration Test (SPT), N values of ~0, Figure 14.4. Their origin or formation plus additional characteristics have been discussed previously (Tan 1988,

Figure 14.5.   Highly irregular or pinnacled limestone bedrock, Sunway mine/quarry, Subang Jaya, Selangor Malaysia (now Bandar Sunway).

1990a, 2004b), Tan & Ch'ng (1986), Ting (1985). The slumped zone has been encountered in numerous high-rise building sites in Kuala Lumpur, including the KLCC or Petronas Twin Towers site, Hamdan & Tarique (1995). Typically for the Kuala Lumpur area, the slumped zone comprises collapsed materials of the Kenny Hill formation that overlies the Kuala Lumpur Limestone. The thickness of this slumped zone can vary from several meters to tens of meters. The slumped zone can also occur at great depths, say ~100 m from the ground surface. Note that overlying the slumped zone, materials representing the original, undisturbed residual soils of the Kenny Hill formation can have high SPT N values of, say 30–50, or even >50, i.e. stiff to hard materials. The slumped zone thus represents a hidden danger or soft "bottom" that can pose problems if undetected earlier during site investigation.

Figure 14.5 shows an example of the highly pinnacled limestone bedrock exposed in a previous tin mine/ex-quarry in the Sungai Way area (now Bandar Sunway), Subang Jaya, a suburb of Kuala Lumpur. Similar exposures of karstic limestone bedrock are retained at the Sunway Lagoon area as they enhance the asthetics of this recreational site.

Foundations in residual soils/rocks

Residual soils from various rock formations, except limestone, show increasing SPT values with depth, hence increasing stiffness/bearing capacity/shear strength with depth. As such, foundations in residual soils, whether shallow footings or deep piles, are not an issue in general. Numerous S.I. data all over the country show this typical trend of SPT increasing with depth for granitic soils, residual soils of shale, schist, etc. This is hardly surprising since the grade of weathering varies with depth in a systematic fashion in general. Exceptions would be where deep fault/shear/brecciated zones are encountered, highly weathered sills and dykes are encountered, in which case this straight forward trend might not be followed. Basic and ultra-basic dykes such as dolerite dykes are commonly encountered intruding phyllite or schist formations in Terengganu, for example, and they tend to be much more weathered than the surrounding phyllite/schist and hence form weaker zones problematic as slope or foundation materials. Quartz dykes, on the other hand, always occur as much stronger intrusions in weaker country rocks and are less problematic. However, quartz dykes tend to be more fractured and can serve as conduits for groundwater flow/seepage, in which case they can contribute to slope instability, tunnel seepages, dam leakages, etc.

A word of caution on the residual soils of graphitic schists and carbonaceous shales which often contain pyrite which produces acidic groundwater/discharges on oxidation and hydrolysis. Foundation soils and fill materials comprising graphitic schist soils can thus cause corrosion and ugly staining of concrete structures, such as footings, drains and pavements. Examples of such incidence can be found in the Bukit Beruntung (north of Kuala Lumpur) area where graphitic schist soils are abundant and have been used as foundation and fill materials for some housing projects. The graphitic schist soils also have high carbon/graphite contents, hence they will yield lower compacted density and shear strength/bearing capacity. In addition, due to the continuous chemical changes involving the oxidation and hydrolysis of the pyrite, there is prolonged disintegration of the soil structure and deterioration in shear strength with time. Thus, graphitic schist soils are not suitable as foundation or fill materials for buildings, embankments, etc.

Dams and damsite investigations

Of all the engineering works, dams are perhaps the most interesting and challenging for the engineering geologist. Damsite investigations not only cover the main damsite foundations; they also include other possible saddle dams, diversion tunnels, spillways, dam abutments, and the entire reservoir and catchment areas. Other investigations would include material survey for rock quarries, earth/clay borrow pits, sources for construction sands, etc. The site and regional geologic settings, including seismic risk, reservoir induced seismicity (in the larger dams), potential major fault movements/reactivation, etc. have to be dealt with. Faults are particularly relevant since dams are built across rivers, and often rivers flow along fracture traces or fault lines. Thus, possible problems from reactivation of fault movement, seepage losses, need for grouting of dam foundation, permeability tests, etc. are matters of great concern in the investigation of damsites. The type of dam selected can also be dictated by the availability of materials: for example, rockfill versus earth dam depending on the availability of suitable rock or soil at close proximity to the damsite.

A review of some local experiences in damsite investigations and related issues can be referred to in Tan (1995c). Details of individual case studies on dams are contained in Tan & Wong (1982), Au Yong & Tan (1984), Tan (1983a, 1994c and 1994d). In the case of the Tawau dam in Sabah, the volcanics in the area (andesite and basalt) were mostly weathered to thick layers of dark brown silty soils, with only minor, isolated occurrences of fresh basalt (with its typical polygonal, columnar jointing) at some distance upstream from the damsite. Though three small hills representing stocks of an intrusive, igneous rock (diorite) occur near the damsite, due to environmental and other landuse considerations, the rock has to be procured from the hill located furthest away from the damsite, hence increasing the cost of the project. In a recent (1999) investigation of a proposed damsite in Sarawak, the dam type selected is likely to be changed from the originally proposed rockfill dam to an earth dam due to the lack of suitable rock source not too distant from the damsite. Incidentally, the use of a helicopter to fly over the damsite and reservoir area can be very fruitful and cost-effective, since ground survey can be hampered by poor accessibility of forested areas associated with dam projects.

14.4 ENGINEERING GEOLOGY IN RELATION TO ROCK SLOPE ENGINEERING

Rockfall in limestone cliffs

The stability of limestone cliffs is one of the major concerns for development projects near or adjacent to limestone hills, such as around the Batu Caves area in Kuala Lumpur, and the Ipoh and vicinity areas in the Kinta Valley where numerous limestone hills are encountered.

Rockfalls represent one of the major geohazards associated with limestone cliffs, and their past occurrences have been documented from time to time, for example, Shu & Lai (1980), Shu et al. (1981), Chow & Abdul Majid (1999), etc.

A major rockfall in 1973 at Gunung Cheroh, Ipoh, which inflicted 40 human fatalities and numerous cows as well, was documented by Shu & Lai (1980). This rockfall involved the collapse of the entire cliff face as a single slab measuring some 33 m in length and weighing 23,000 tonnes and was one of the major rockfall disasters associated with limestone hill in the country. The corpses entombed have never been recovered due to the enormous size of the rock slab involved.

The engineering geology of limestone, covering both limestone hills and bedrock, with their associated problems such as rockfalls and sinkholes, cavities, slumped zone, etc. has been reviewed by the author in a recent paper, Tan (1999c). Further details can be referred to in the IEM-GSM forum proceedings on karst/limestone mentioned above. Of particular interest and reproduced here is the case study on the survey of limestone cliffs for urban development purposes in the Tambun (Ipoh) area. The limestone cliffs were surveyed for geological structures (fractures) and solution features, and were then assessed for their stability based on the combination of these features. Hazard maps were then produced to indicate various segments of the cliff faces with differing stability or degree of hazards. Table 14.1 shows some sample results of the survey conducted, and

Table 14.1.   Summary of field survey data, Tambun limestone hills.

Segment	$\theta°$/H (m) (θ = slope angle) (H = slope height)	Structure J/B/F (J = joint, B = bedding, F = fault)	Solution features	Rockfalls
1 to 2 (Hot spring)	80° (O)/60 m (O) = overhang	J1 95/75 (O) J1 94/78 (O) J1 92/76 (O) J2 100/77 J2 101/70 J2 125/76	Cave φ(160°) d = 10 m l = 50 m (diameter = d, length = l) φ(160°) = cave axis	Nil
2 to 3	90° (V)/60–150 m (V = vertical)	Vegetation (Vegetation = Veg.)	Cavity d = 2 m l = 2 m	Nil
3 to 4 (Temple to pool)	90° (V)/50 m	J1 60/80 J1 54/74 J1 60/80 Vegetation in cliff (end of segment 4)	Cave φ(75°) d = 20 m l = 20 m Solution slot along J1 Cave φ(100°) d = 2 m l = 5 m	Nil
4 to 5	90° (V)/150–80 m	J1 52/65 J1 64/69	Nil	Nil
5 to 6	90° (V)/80 m	J1* 105/88 J1* 107/84 J1 10/82 J1 11/72 J1 15/68 J2 118/69 J2 115/76 J2 120/72	Nil Nil Nil	Nil Nil Nil
6 to 7	90° (V)/50 m	Vegetation	Nil	Nil
7 to 8	80°–90° (V)/50 m	J1 115/70	Nil	Nil
9 to 10	90° (V)/60 m	B 060/45	Cave φ(30°) d = 2 m l = 2 m Undercut & Stalactite	Nil
11 to 12	90° (V)/150 m	Vegetation	Nil	Nil
12 to 13	90° (V)/150 m	J1 175/54	2 small cavities	Rockfalls
14 to 15	90° (V)/100 m	Vegetation	Nil	Nil

Figure 14.6 shows an example of the hazard map produced for some segments of the limestone cliffs. The results of the survey indicate that the stability of the various limestone cliffs was controlled by-and-large by the north-south trending bedding planes. More details on this case study are contained in Tan (1998b).

Figure 14.7 shows a very recent (2005) case study on the assessment of limestone cliff stability adjacent to a water tank site in Batu Caves, Kuala Lumpur. The stability of the limestone cliff is

Figure 14.6.   Example of hazard zonations of limestone hills (RF = rockfall), Tambun, Perak Malaysia.

Figure 14.7.   Water tower adjacent to limestone cliff, Batu Caves, Kuala Lumpur Malaysia.

of great concern, and again, assessment is based on the combination of geologic structures and solution features present at the limestone cliff in question.

Urban geology and hillside development

Urban geology and hillside developments are major topics for the engineering geologist. As the name implies, urban geology is the application of geology to urban development, as amply illustrated in the book by Legget (1973). The case of the collapse of the Highland Tower in 1993 also brought to centre stage the problems associated with hillside development, not only in Kuala Lumpur but also in other urban centres in Malaysia. Useful papers related to engineering geology and geotechnical engineering in hillside development are contained in the Proceedings of the Symposium on Hillside Development: Engineering Practice and Local By-Laws (1995) published by IEM. Geologic input and engineering geology in urban and hillside development have been discussed previously, Tan (1986a, 1986b, 1990a, 1990b, 1993 and 1995b), Tan & Komoo (1990). In addition to the problem of limestone bedrock and limestone cliffs for the Kuala Lumpur and Ipoh areas, other urban and hillside development problems include landslides in its widest definition, involving both rock and soil materials, including man-made fill, mine tailings, colluvium, etc. The fundamentals of engineering geology, namely: lithology (material type), structure, and weathering are needed input in the various studies on urban and hillside development, even in the case of simple housing projects located near hill slopes.

Figure 14.8 provides some examples of engineering geologic studies for housing development projects located in hilly terrains in the Kuala Lumpur area. Survey of rock slopes and assessments of potential problems are made to aid in the planning and implementation of slope stabilization measures.

Engineering geology of highways and roadways

Highways and roadways require significant engineering geological input. Besides mapping the highway route geology so that the rock and soil types traversed by the highway can be known in advance, geological input during the construction stage are also very necessary. Excavation and treatment of cut slopes, sourcing for construction materials such as rock quarries, earth borrow pits, etc. require engineering geological input. The various works involved in the geological and hydrogeological investigations for highways and roadways have been summarized in Tan (1994a). Example case studies on the engineering geology of highways and roadways are presented in Tan (1987, 1992 and 1994b) which include the Karak Highway, parts of the North-South Expressway, and a rural road in Sarawak.

Detailed mapping of cut slopes along an existing highway has become a routine work for the engineering geologist involved in the assessment of cut slope stability and in post-failure investigations for remedial works. Figure 14.9 shows a sample field survey sheet for cut slopes for such purposes. Note that the three fundamental items in engineering geology, namely lithology, structures and weathering form the gist of the survey sheet. Assessment of slope stability can then be carried out using the stereonet method of Hoek & Bray (1974). The data and sketches on weathering profiles can also help in slope treatment or remedial works, such as proportioning rock and soil slopes for different treatment works. Hydroseeding, for example, is only suitable for soil slopes, i.e. grades IV to VI materials only. Blindly spraying the entire cut slope without due regard to the weathering grades/profiles is wasteful, to say the least.

Table 14.2 and Figure 14.10 provide sample results of the engineering geological surveys conducted for the Senawang–Air Keroh Highway. Route geology depicted in Figure 14.10 for the Senawang–Air Keroh Highway shows the various rock formations traversed by the highway, in particular the predominance of graphitic schist towards the Air Keroh end of the highway. The survey results show that the majority of cut slope failures occurred in the graphitic schist material, indicating geological or material control in the failures. In the case of a rural road in Sarawak, shale slopes are the problematic slopes compared to the sandstone slopes; in addition, structures such as bedding planes/foliations and faults, produce the major slope failures.

Stereoplots of Rock Joints	Grade of Weathering	Potential Failure Mode	Remedial Measures	Remarks
	II mainly, some III-IV	J2, J1: toppling J3: planar(10-30°) J1-J3: wedge (20°)	Major rock bolting	Major toppling hazard, adjacent to condominium
	I/II, some III	J3: toppling J4: planar (>30°) J1-J4: wedge (45°)	Generally stable, minor spot bolting and dentition	Slope height 3m only, joint spacing 1/2-2m.
	II/III mainly, some I	J3: toppling J1-J4: wedge (60°)	General cleaning & scaling of loose blocks. Rock fenders or wire nets. Some minor spot bolting.	Crucial since slope is behind(1-3m) completed house. Rock failures include cavities & fallen blocks. Seepage along sheet joints.
	I/II, some III	a) J3: planar J1-J3: wedge (14°) J2-J3: wedge (14°) b) J2: planar c) J3: rocfalls/ wedges.	Benching by controlled blasting(presplit, etc) Trim blasting to remove overhangs, followed by cleaning & scaling, rock fenders/wire nets & spot bolting.	Slope height = 20m. Excessive overhangs, perched loose blocks. Excessive seepage. Previous planar slide.
	I/II, some III	a) J2: toppling b) J3: toppling c) J3: toppling	Minor spot bolting. Concrete rock buttress for major overhang. Dentition or rock buttress for major overbreak.	Major overhang & over-break. 2 major faults. Height = 2-3m only for b).
	I/II mainly, some III - VI	a) J1: planar J1-J3: wedge (30°) J1-J2: wedge (40°) b&c) J1: toppling J2, J3: planar J2-J3: wedge (40°) J2-J4: wedge (40°) J3-J4: wedge (36°)	Controlled blasting, followed by major rock bolting program.	Height = 20m. Full of instability problems. (overhangs, perched slabs, soil-rock debris slides, etc)

Figure 14.8.   Summary of granite rock slope conditions and possible remedial measures, housing scheme, Kuala Lumpur, Malaysia.

Figure 14.11 shows a classic, textbook example of a wedge failure in granite rock slope for the newly completed SILK highway in Kajang. Good examples of major plane failures along major daylighting joints in the granite rock slope are also found along this highway. Detailed mapping of the various major joint sets in the granite rock mass along the highway is necessary for rock slope stability analysis and implementation of remedial measures such as rock bolting, etc.

A recent major cut-slope failure involving graphitic schist at the Lojing Highway is shown in Figure 14.12. Slope failure occurred at the same spot here at least three times, twice after re-grading of the slope, indicating again the weak nature of the graphitic schist soils. These repeated failures of the same slope are reminiscent of similar failures involving graphitic schist along the Senawang–Air Keroh Highway, and again clearly illustrates the very weak nature of the graphitic schist soils. Another major slope failure involving the graphitic schist soils occurred along the North-South Highway near Rawang – again, slope failure occurred repeatedly even after re-grading. The slope has since been practically "flattened".

Tunnels

Tunnels and tunnelling are rather specialised engineering projects that require advanced technology and know-how. Investigations for tunnels are targeted towards the determination and classification

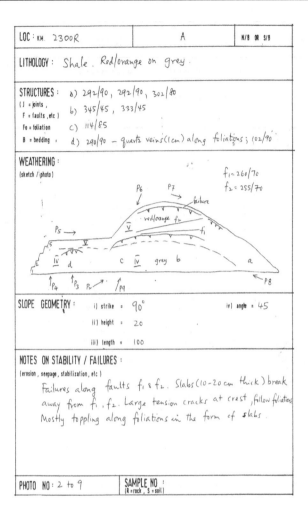

Figure 14.9.   Field survey sheet/check list, Sarawak rural road.

Table 14.2.   Summary of cut slopes surveyed, Senawang–Air Keroh Highway.

	A		B		C	
	No.	%	No.	%	No.	%
Graphitic Schist	8	66.7	11	44.0	7	9.3
Quartz-mica Schist	3	25.0	8	32.0	15	20.0
Quartzite/Phyllite	1	8.3	6	24.0	45	60.0
Granite	–	–	–	–	8	10.7
Total	12		25		75	

of the quality of the rock masses along the tunnel alignment, for the purpose of assigning the type of tunnel support or support systems along the various segments of the tunnel. The method of tunnelling adopted, for example drill-and-blast (New Austrian Tunnelling Method, NATM) versus tunnel boring machine (TBM), would also depend on the type and quality of the rock masses. Tunnel investigations involve drilling along the tunnel alignment prior to construction, and then detailed

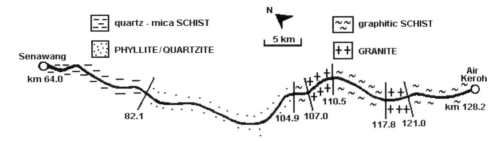

Figure 14.10.   Highway route geology, Senawang–Air Keroh Highway.

Figure 14.11.   Classic, textbook example of wedge failure in granite rock slope, SILK highway.

Figure 14.12.   Slope failure in graphitic schist, Lojing Highway.

Figure 14.13. Heavy steel sets support for highly fractured/sheared granite, Sungai Selangor Dam diversion tunnel.

Figure 14.14. Guniting and bolting of weathered granite rock slope at tunnel portal, Sungai Selangor Dam diversion tunnel.

mapping of geology and fractures/discontinuities during tunnelling. Various classification systems in relation to tunnel support and rock quality have been proposed, the Norwegian system being the most widely used and accepted worldwide. Details of the Norwegian Q-system are contained in the Proceedings of Seminar on Tunnelling in Weak Rocks, June 1999, Kuala Lumpur and also in Barton et al. (1974), Kveldsvik & Karlsrud (1995).

Figure 14.13 shows heavy steel sets support for the highly fractured and sheared granite in the diversion tunnel of the Sungai Selangor dam at Kuala Kubu Baru. The heavily reinforced slope of the portal area is shown in Figure 14.14, including shotcreting and rock bolts/rock dowels. A country report on tunneling activities in Malaysia was presented by Ting et al. (1995).

Table 14.3 summarizes some tunneling activities and the geological factors involved. Some earlier case studies can also be found in Tan (1983b). In the diversion tunnel of the Upper Muar dam, numerous vertical faults were not detected during the site investigation stage which utilized only a limited number of boreholes, all vertical (no inclined boreholes). During the excavation of the tunnel by the drill-and-blast method, frequent collapses and excessive overbreaks occurred when the tunnel intersected the vertical faults with their crushed and weathered materials. The excessive overbreaks led to extra concreting for the tunnel linings, hence claims for additional works and some disputes. This simple example illustrates the need for adequate and suitably conducted boreholes (number plus inclined boreholes if faults are vertical).

Table 14.3.   Some tunneling activities in Peninsular Malaysia Ting et al. (1995).

Item	Name of project	Application	Geology	Observation
(i)	Upper Muar Dam (Water supply), Negeri Sembilan	Diversion tunnel	Granite/faulting	Excessive over-breaks & collapses due to faulting.
(ii)	Kenyir Dam (Hydro-electric), Terengganu	Diversion tunnel	Granite/jointing	Sub-horizontal sheet joints required bolting.
		Pressure tunnel	Granite/jointing	Sub-vertical joints at portals required close-grid bolting.
(iii)	Batang Padang (Hydro-electric) Scheme, Pahang	Diversion tunnel Pressure tunnel Rock cavern (Power house)	Granite/faulting/ weathering	Tunnels collapsed (soil flows) due to tunneling through weathered materials associated with fault. In-situ ground stress measurements + rock mechanic studies.
(iv)	Sg. Piah (Hydro-electric) Scheme, Pahang	Diversion tunnel Pressure tunnel	Granite/faulting + Schist (roof pendant)	Relocation of tunnel portals away from major fault.
(v)	Pergau Dam (Hydro-electric), Kelantan	Diversion tunnel Pressure tunnel	Granite/gneiss	On going.
(vi)	Kelinchi Dam Water supply, Negeri Sembilan	Transfer tunnel	Granite/jointed/some faults anticipated	On going. Only tunnel excavated using Tunnel Boring Machine.
(vii)	Ahning Dam (Water supply), Kedah	Tunnel for water conduit	Sandstone/shale/ conglomerate	–
(viii)	Pedu/Muda Dams (Water supply), Kedah	Transfer tunnel	Sandstone/shale/ conglomerate	–
(ix)	Karak Highway, Pahang	Highway tunnel (single)	Granite/jointing	Additional tunnel being constructed beside existing one.
(x)	Changkat Jering Highway, Perak	Highway tunnel (twin)	Granite/jointing	Massive/stable.
(xi)	Batu Arang Coal Mine, Selangor	Underground mine/tunnel/adits (abandoned)	Coal/sandstone/ shale	Sinkholes & subsidence due to tunnel collapses.
(xii)	Sg. Lembing tin mine, Pahang	Underground tin mine/numerous tunnels/adits (abandoned)	Granite/quartz dykes intruding metasedimentary rock	Seepage/pumping problems due to deep mines.
(xiii)	Kaki Bukit tin mine, Perlis	Tin mine along solution channels/ tunnels (abandoned)	Limestone hill/ bedrock	Mining alluvial tin along solution channels in limestone.
(xiv)	Hanjung Cement Plant, Perak	Access ramp/ road tunnel	Limestone hill	Quarrying for limestone. Caves/stalactites etc. along tunnel route.
(xv)	Ammunition Depot, Tg. Gelang, Kuantan, Pahang	Storage tunnels	Phyllitte/Quartzite, complex folding & faults	Heavy tunnel supports steel arches + resin grouted rock bolts.

14.5   CONCLUSIONS

Engineering geology is, needless to say, important in various engineering construction projects. Engineering geology in relation to foundation and rock slope engineering are discussed in this chapter, and they cover topics on limestone, residual soils, urban and hillside development, highways and roadways, dams and tunnels. Fundamental input of engineering geology in many engineering projects would include studies on lithology, structure and weathering grade, as these factors control the behaviour of the rock mass. Numerous references are cited and listed below for the reader who may wish to seek further details on some of the case histories mentioned.

ACKNOWLEDGEMENTS

The author acknowledges numerous friends in the civil and geotechnical engineering fraternity who have provided the opportunities for the involvement in various engineering projects, hence to practise engineering geology, in particular Dr. Ting, W.H., Dr. Chan, S.F. & Dr. Ooi, T.A.

REFERENCES

Au Yong, M.H. & Tan, B.K. 1984. Construction materials for the Sembrong and Bekok Dams, Johor. *Bulletin Geological Soc. Malaysia*. (17, 1994): 49–60.

Barton, N., Lien, R. & Lunde, J. 1974. Engineering classification of rock masses for the design of tunnel support. *Rock Mechanics*, 6/4: 189–236, Vienna: Springer Verlag.

Chow, W.S. & Abdul Majid Sahat 1999. Rockfalls in limestone hills in the Kinta Valley, Perak, Malaysia. IEM-GSM Forum on Karst: Geology & Engineering, Petaling Jaya, Paper (3): 19.

Hamdan Mohamad & Tarique Azam 1995. The Petronas Towers, the tallest building in the world. *Journal Institution of Engineers Malaysia* 56(1, April 1995): 119–134.

Hoek, E. & Bray, J.W. 1974. *Rock slope engineering*. Institution of Mining & Metallurgy, London: 309.

ISRM 1977. Suggested methods for the quantitative description of discontinuities in rock masses. *Int. Soc. for Rock Mechanics, Commission on Standardization of Laboratory and Field Tests*, Committee on Field Test, Doc. Oct. 1977, (4): 319–368.

Kveldsvik, V. & Karlsrud, K. 1995. *Support methods and groundwater control*. Norwegian Urban Tunnelling, Norwegian soil and Rock Engineering Association, Tapir Publishers, Univ. of Trondheim, Trondheim, (10): 69–77.

Legget, R.F. 1973. *Cities and geology*. New York: McGraw Hill.

Legget, R.F. & Karrow, P.F. 1983. *Handbook of geology in civil engineering*. New York: McGraw Hill.

Little, A.L. 1969. The engineering classification of residual tropical soils. *Proc. Specialty Session on the Engineering Properties of Lateritic Soil, Vol. 1, 7th Int. Conf. Soil Mechanics & Foundation Engineering*, Mexico City 1: 1–10.

Shu, Y.K., Chow, W.S. & Zakaria, M. 1981. Rockfall danger related to limestone hills in the Kinta Valley, Perak. Annual Report, Geological Survey Malaysia: 184–197.

Shu, Y.K. & Lai, K.H. 1980. Rockfall at Gunung Cheroh, Ipoh. Geological Survey Malaysia, Geological Papers 3: 1–9.

Tan, B.K. 1982. Engineering geology. Chapter 3, Short course in geotechnical engineering, Univ. Malaya/IEM, 22 March-2 April, Kuala Lumpur: 40.

Tan, B.K. 1983a. Geotechnical aspects of the Kenyir Dam project, Trengganu, Peninsular Malaysia. *Proc. 5th Int. Congress on Rock Mechanics, ISRM*, 10–15 April 1983, Melbourne: C133-C137.

Tan, B.K. 1983b. Engineering geological case studies of tunnelling and deep excavations in Malaysia. *Proc. Int. Symp. on Engineering Geology and Underground Constructions*, 12–15 Sept. 1983, Lisbon, I.25-I.33.

Tan, B.K. 1986a. Geology and urban development of Kuala Lumpur, Malaysia. *Proc. LANDPLAN III Symp.*, 15–20 Dec. 1986, Hongkong, 127–140. (Geological Society of Hongkong Bulletin (3), Oct. 1987).

Tan, B.K. 1986b. Landslides and hillside development – recent case studies in Kuala Lumpur Malaysia. *Proc. LANDPLAN III Symp.*, 15–20 Dec 1986, Hongkong, 373–382.

Tan, B.K. 1987. Engineering geological studies on landslides along the Kuala Lumpur-Karak Highway, Malaysia. *Proc. IAEG Conf.*, Beijing China, 4–8 May 1987, 1: 347–356.

Tan, B.K. 1988. A short note on the occurrence of a soft soil zone above limestone bedrock. *Proc. Int. Conf. on Calcareous Sediments*, Perth, March 15–18, 1988, 1: 35–39.

Tan, B.K. 1990a. Subsurface geology of Ipoh area, Perak, Malaysia. *Proc. Conf. Karst Geology in Hongkong*, 5–7 Jan. 1990, Hongkong. (Geological Society of Hongkong Bulletin (4), 1997): 155–166.

Tan, B.K. 1990b. Engineering geology of Ipoh, Perak, Malaysia. *Proc. 6th IAEG Congress*, 6–10 Aug. 1990, Amsterdam, 3: 1733–1738.

Tan, B.K. 1991. Role of geology in geotechnical engineering. 4-Day Course on Geotechnical engineering, IEM, 9–12 Sept 1991, Petaling Jaya, 3: 28.

Tan, B.K. 1992. A survey of slope failures along the Senawang–Air Keroh Highway, Negeri Sembilan/Melaka, Malaysia. *Proc. 6th Int. Symp. on Landslides*, 10–14 Feb. 1992, Christchurch, 1423–1427.

Tan, B.K. 1993. Urban geology of Ipoh and Kuala Lumpur. *Proc. Forum on Urban Geology and Geotechnical Engineering in Construction, IEM/GSM*, 1 July 1993, Petaling Jaya: 1–1 to 1–25.

Tan, B.K. 1994a. Geological/Hydrogeological investigations for highways and roadways. Short Course on Engineering Geology for Civil Engineers, IKRAM, 18–23 July 1994, Kajang, 11pp + 2 Appendices.

Tan, B.K. 1994b. Survey of slope failures for a rural road in Sarawak. Newsletter, Geological Soc. Malaysia, v.20, n.4, July-Aug 1994, 285–290.

Tan, B.K. 1994c. Investigations for the Gemencheh Dam, Negeri Sembilan. *Proc. GEOTROPIKA 94*, Malacca, 22–24 Aug 1994, Paper 3–3.

Tan, B.K. 1994d. Investigations for the Tawau Dam, Sabah, Malaysia. *Proc. 7th IAEG Congress*, Lisbon, Portugal, 5–9 Sept. 1994, V: 3707–3714.

Tan, B.K. 1995a. Some experiences on weathering of rocks and its engineering significance in Malaysia. IKRAM Geotechnical Meeting 1995, 7–9 June 1995, Penang, 2 (Lect.6): 22.

Tan, B.K. 1995b. Geologic input in hillside development – some case studies in Malaysia. *Proc. IEM Symp. On Hillside Development: Engineering Practice & Local By-Laws*, 5–6 June 1995, Petaling Jaya, Paper 8: 16.

Tan, B.K. 1995c. Damsite investigations – Malaysian case studies. *Proc. Int. Conf. on Dam Engineering '95, 1–2 Aug 1995, Kuala Lumpur*, Malaysian Water Association & Canadian Dam Safety Association 457–462.

Tan, B.K. 1998a. Checklist for site investigation work – geology/engineering geology. Short Course on Geotechnical Engineering, IEM, 17–18 June 1998, Petaling Jaya: 63.

Tan, B.K. 1998b. Assessments and hazard zonations of limestone cliffs in the Tambun area, Perak, Malaysia. *Proc. Regional Symp. on Sedimentary Rock Engineering*, Taipei, Nov. 20–22, 1998: 55–59.

Tan, B.K. 1999a. Engineering Geology – some case histories in Malaysia. Keynote Paper. *Proc. 5th Geotechnical Engineering Conf. (Geotropika 99)*, UTM, Johor Baru: 23–43.

Tan, B.K. 1999b. Basic mechanics of slope failures & geological investigation of potential failure modes. IEM short course on rock slope engineering, 23 March 1999, Petaling Jaya.

Tan, B.K. 1999c. Engineering geology of limestone – an overview. IEM/GSM Forum on Karst: Geology & Engineering, 26th Aug. 1999, Petaling Jaya.

Tan, B.K. 2004a. The practice of engineering geology in Malaysia. *Special Lecture, Proc. Malaysian Geotech. Conf. 2004*, March 2004, Subang Jaya, 131–148.

Tan, B.K. 2004b. Country case study: engineering geology of tropical residual soils in Malaysia. *Proc. Symp. on Tropical Residual Soil Engineering – TRSE2004*, 6–7 July 2004, University Putra Malaysia, Serdang, Invited Lecture, Chapter 14, 237–244, Balkema.

Tan, B.K. 2004c. Engineering geology of rock slopes – some recent case studies. *Proc. GSM-IEM Forum on the Roles of Engineering Geology & Geotechnical Engineering in Construction Works*, 21st Oct. 2004, Kuala Lumpur (8): 11.

Tan, B.K. & Ch'ng, S.C. 1986. Zon tanah lemah di atas batu kapur dasar (Weak soil zone above Limestone bedrock). Newsletter, Geological Society of Malaysia, Mar–April 1986. 12(2): 51–57.

Tan, B.K. & Ibrahim Komoo 1990. Urban geology: Case study of Kuala Lumpur, Malaysia. Engineering Geology, 28(1990): 71–94.

Tan, B.K. & Wong, P.Y. 1982. Site investigation for the Bekok Dam, Johor, Malaysia. *Proc. 4th Int. Congress on Engineering Geology, IAEG*, 10–15 Dec.1982, New Delhi: III.11–III.17.

Tarique Azam 1996. Project KLCC: Geology, soils and foundations. Newsletter, Geol. Soc. Malaysia, March–April, 1996. 22(2): 73–74.

Ting, W.H. 1985. Foundation in limestone areas of Malaysia. Special Lecture, *Proc. 8th S.E. Asian Geotechnical Conf.*, Kuala Lumpur, March 1985, 2: 124–136.

Ting, W.H. & Nithiaraj, R. 1998. The stabilization of a slope in weathered formations in Malaysia by ground anchorages. *Proc.13th S.E. Asian Geotech. Conf.*, Taipei, Nov.16–20 1998, pre-print 1: 44–55.

Ting, W.H., Ooi, T.A. & Tan, B.K. 1995. Tunnelling activities in Malaysia – Country Report. *Proc. S.E. Asian Symp. on Tunnelling & Underground Space Development*, 18–19 Jan 1995, Bangkok.

Author Index

Subject Index

T - #0054 - 071024 - C0 - 254/178/13 [15] - CB - 9780415398985 - Gloss Lamination